INTEGRITY DATA PROTECTION FORENSIC

[COMPUTER FORENSIC TECHNOLOGY]

NEW TREND

INTEGRITY DATA PROTECTION FORENSIC

[COMPUTER FORENSIC TECHNOLOGY]
NEW TREND

MARIO NABLIBA

To order additional copies of this book, contact:

Xlibris Corporation

1-888-795-4274

www.Xlibris.com

Orders@Xlibris.com

119343

CONTENTS

New Trend

Understanding science is the joy of our life endeavors.

This phrase leads me to quote with my humble idea, which is derived and inspired from a quote of Albert Einstein:

> How many times a day I remind myself that my life depends on the labors of other men—great inventors, scientists living and dead—for the better of this nation, and that I must exert myself in order to give, in the measure as I have received so much on America soil, and I am still receiving. *(Mario Nabliba as reflected by Albert Einstein in this New Trend book)*

Acknowledgments

I would like to thank the many people who have helped me at various stages in my research of this book during the past years. Their encouragement, criticism, and suggestions were invaluable to me.

A special note of thanks goes to the US Department of Education and the federal government for an unsubsidized loan on my graduate study program. Without their help, I would not have been able to arrive at these amazing conclusions of IDP Forensic. To my lovely wife, Suzanne Elizabeth Nabliba "Crowley," without whom I could not have completed this work. My brother-in-law, Tim Crowley, for his help; Dr. Jose Ligna Na Fafe, professor at Birmingham University of England, for providing support logistically on my research trip; and Dr. Silva of the faculty of the Science, University of Lisbon, who many times talked about the impact of the Renaissance on future generations to come. My friends, I thank you all. I would like to thank my mom and dad because they accept and love me as part of them. Again, I want to thank my wife for her understanding in letting me work such long hours in the career of a scientist that I love and that this is possible in the greatest nation in the world—the USA.

Chapter 1

IDP FORENSIC NEW TREND ENTERPRISE BUSINESS PLAN WILL AVOID A FINANCIAL HARDSHIP

To control our own performance today, many organizations require more attention to the success of each individual in business, particularly entrepreneurs and big corporations.

- Predict the situations or outcomes of the behavior of individuals.
- Provide the individual with innovation and training for the use of outstanding technology.
- Think well and present the most accurate ideas related to business.
- Orient the individual to his role and how he will behave in functions of management.

IDP FORENSIC NEW TREND ENTERPRISE BUSINESS PLAN

Integrity Data Protection Forensic
4225 Via Arbolada #501
Los Angeles CA 90042
Tel: 323-450-6826
E-mail: info@idpforensic.com
www.IDPForensic.com

American Institute IDPF

INTRODUCTION

IDP Forensic is a field of science, computer investigation, and analysis techniques that involve the identification, preservation, extraction, documentation, and interpretation of computer data to determine potential legal evidence.

MISSION STATEMENT

Integrity Data Protection Forensic is a company dedicated to the virtue of intellectual properties.

The mission of Integrity Data Protection Forensic LLC is to serve any field with outstanding performance in the area of computer forensic science—professional presentation of evidence in a court of law. Our organization works with outstanding, high-tech legal agents who use technology without compromise of integrity. Therefore, our product is meant to further our client's prosperity in regard to the protection of their intellectual properties.

We are professionals trained to use hardware and EnCase software and to investigate and uncover the most hidden pieces of information that will resolve a client's issue. If a crime happens, our core services in computer forensics will present in detail each digital evidence found. We stand behind our work.

According to our goal and objective, we have achieved key success by sending personnel for training. Our "competitors" are our friends and are part of our social network, so we are actually working as a team. As an example, Guidance Software is a part of our team who invited us for conferences.

As the success of IDP Forensic will depend on conviction and hard work, IDP's plan is to develop the company into the American Institute of Integrity Data Protection Forensic.

MARKETING

Our clients are those who have intellectual properties to protect. In this high-tech age, most companies will need our help at one time or another. To start this organization, we want to begin with hiring citizens with high integrity and who are the most patriotic members of America. This is a strong drive with the founding members of this company. We know, with unemployment, we must help our economy and especially our returning US citizens.

Our project seeks to enhance education with a Department of Translations, which will take care of official translations of instructional manuals and official papers such as programming languages. from foreign languages to English and vice-versa

Most of the project falls into three main categories:
Department of Academic Research and Discovery—this department is designed to follow the goal of American invention and innovation. To work in this department requires knowledge in the field of computer science and a variety of study and work experience. Having these achievements will qualify the scientists to be able to demonstrate knowledge and to explore the real world by using work experience. This requirement is based on an academic curriculum of a university of the 21st century.

To do research and discovery, you must consider the research as part of our culture and our axioms: an achievement that is a must for each individual is to discover a new American trend. Each qualified individual must be able to make discoveries in his area of focus according to US standards and regulations. Therefore, we will work closely with other national organizations, such as NST (National Science Technology).

Locally, our partners are enthusiastic about publicizing our project: first, to let our government know about this project and, secondly, to gain support for services. All partners have agreed to publicize the project through their newsletters, and so you are welcome to be a part of this great company, IDP Forensic. But more importantly, we will each make key contacts with a legal agent. We will work with the city governments and federal governments to coordinate these efforts. A mailing list is to be compiled from public records and from local agencies. A letter with a brochure will be mailed to each person. Costs will be covered by a combination of partners through their public relations offices.

South Pasadena Police are supporting this project, and they recognize our organization fills a need in the community.

Nationally, we will work through the National Association. Our county commissioner is active with the association and will help us make the right contacts to get publicity. We understand that a new company will have challenges, so we are setting up our approach with an internal weekly audit of the funds provided to us to ensure their proper use. Our grantor can request at any time to be informed of the use of the money. The name of grant donors will be recognized and listed with great honor.

In addition, we will join as members of all relevant associations nationally and actively pursue publicity for our project. This will be done by submitting articles to newsletters, participating in conferences as speakers and attendees, and seeking reciprocal links to websites.

FINANCES

Our capital equipment and supplies list is the following:

1Dell laptop 13"	$1,300
+	
80 gigabytes storage data	$100
+	
Legal Business Pro 2009	$40
+	
Pro Office System	$45
+	
Website	$1,000
+	
Training Staff (CEIC)	$2,000
Total	$4,485

There are many other resources the business needs to be successful, including a building to operate out of. Currently, we have a home-loft office, which is 150 square feet.

Here are some of our products for sale and also that we will be servicing:

Therefore, we are requesting a grant to be able to acquire these resources. When we get the grant, we will be able to achieve the most sophisticated information system technology resources to put into practice the goal of this organization and buy our own facility or office here in the United States. We are requesting this value in dollars as well as in material. We would like a grant for $ . . . , ...00.

MANAGEMENT

Our team is made up of the following professionals:

Mario Nabliba, founder

- Master in computer science information systems, certified in computer forensic science
- Computer system analyst with extensive experience in electronic and computer science
- Bachelor of science in physics
- Expert in project management

Suzanne Nabliba, president

- Thirty years experience in business administration
- Vice President
- BA Business Administration
- Extensive experience in marketing

Tim Crowley, product manager

- Veteran of Vietnam War
- Twenty years' desktop publishing and design management experience.

This is the team—very involved—dedicating their free time to get this project going, and they are realistically doing it.

CONCLUSION

The current status and need of IDP Forensic to be a functional organization is to get a building for IDP Forensic, hire our returning veterans, and acquire resources to sell.

Our goal is to get this grant to start as soon as possible, and with the team we are going to hire acting accountably, we will be able to perform well to get ROI (return on investment) by the end of 2011. The goal is to double the amount of the grant or more given to us. We are a hardworking team, and we believe the giant American dream can come true. President Franklin said, "Honesty is the first chapter of the book of wisdom." And so all American heroes deserve our honor and jobs we can provide, so together we can build a better America.

Thank you for your prompt action to this grant. We shall look forward to hearing from you.

Sincerely,

Mario Nabliba, MS
Founder

Chapter 2

TRACKING AND MANAGING DIGITAL AGE

Products such as mobile phones and IDPForensic Computer Forensic Technology Analysis

An exciting new addition to the field of forensics is the analysis and recovery of information stored on cell phones by way of SIM cards and other media devices.

Someone described cells phones as the new fingerprint. Critical evidence in mobile devices can make or break a case in a court of law.

We extract this vital evidence from any type of cell phone even if someone has tried to delete the information. This includes text messages, phonebooks, and information of any size. This evidence is then displayed in clear concise reports.

For a free consultation with a digital forensic expert regarding cell phones or PDA forensic analysis call now at 323-450-6826.

What we do is evident in the successes we receive from successful people and the time we save them by protecting and recovering their information. They find their production goes up. They deserve the satisfaction, and happiness and freedom we give them in the use of computer forensic technology.

Here is a brief testimony of one of our recoveries on a blackberry:

Optimum Wellness Medical Group

G. Megan Shields, M.D.   Elizabeth Bartlett, FNP
1030 S. Glendale Ave, # 503
Glendale, CA 91205
Phone 818-547-5400   fax 818-547-3380

To Whom It May Concern:

Sept. 4, 2012

LETTER OF RECOMMENDATION
Mario Nabliba

Recently I had a major problem with my cell phone. I thought I had completely lost all passwords to all my accounts – this was a list of more than 50 user names and passwords. The loss was catastrophic.

I gave the problem to Mario Nabliba and he recovered them for me using his forensic skills. He is a genius. This was much appreciated.

Yours truly,

G. Megan Shields, M.D.

Chapter 3

THE NEW FEDERAL EDISCOVERY RULES: EXPANDING YOUR SPHERE OF INFLUENCE

WRITTEN BY JACK MOLISANI AND DR. JOHNETTE HASSELL

When a lawsuit is brought against a company (or individual) in the United States, the US legal system requires the company (the defendant) to produce information that the party filing the suit (the plaintiff) believes might contain information relevant to the claims in the case. For example, there have been numerous high-profile cases in the media over the past year in which corporate executives were suspected of insider trading. In such cases, the plaintiffs subpoenaed copies of correspondence sent or received by the executives in the months prior to the event to determine if they contained evidence that the executives did indeed have "inside information" and acted on it illegally.

The process by which plaintiffs request such information and defendants produce (or in some cases refuse to produce) the information is called *discovery*. Each side is said "to produce" (deliver) material in response to the other's discovery requests.

The US legal system provides a set of rules that govern how the discovery process should be conducted. This article addresses revisions to the US federal discovery rules that went into effect December 1, 2006, and how they provide technical communicators an unprecedented opportunity to expand their sphere of influence.

Important note: the authors are not legal professionals, nor are we qualified to give legal advice. While this article offers a lay interpretation of US discovery rules and suggests actions you can take, you should check with your company's legal team and work with them to help ensure your company is prepared.

THE "OLD" DISCOVERY RULES

For hundreds, if not thousands, of years, "discoverable" information was pretty much limited to written records such as letters, memos, receipts, etc. The media on which information was recorded may have evolved from clay tablets to papyrus to modern office paper, but the content was still *written* (initially written by hand and, later, also printed).

The invention of computers, however, radically changed how information was stored and transmitted. Instead of interoffice memos, we send e-mails. Accounting ledger papers have been replaced by spreadsheets. Shoeboxes stuffed with smudged and dog-eared index cards have been replaced by customer—and relationship-management systems, enterprise-wide accounting and finance systems, and other complex data storage and retrieval systems.

The information age had arrived, and the old discovery rules needed to be updated.

THE "NEW" DISCOVERY RULES

Since 1938, the US legal system has provided laws for the discovery process. These rules assumed information was produced (delivered) on paper. Several revisions were made over the years, the last being in 2000, yet the rules still contained inadequate provisions for the differences between paper-based information and electronically stored information. After a four-year process, the US federal court system created and published new rules aimed specifically at how to manage the discovery of electronically stored information or *ESI*.

The new discovery rules cover (for the purpose of this article) three main areas:

- The proactive new role attorneys must take in understanding a client's entire ESI
- How and in what form ESI should be produced
- What companies should do in the normal course of business (that is, *before* a legal action is initiated)—or risk fines or other legal sanctions

Types of ESI

Let's discuss different types of ESI that are covered under the new discovery rules. Some types of ESI that might immediately come to mind include the following:

- E-mail (and e-mail attachments)
- MS Office files
- Files published to PDF or HTML
- MS Outlook calendars
- Software source files
- Documentation source files
- Website source files

Other types of ESI that might not immediately come to mind include the following:

- Digital audio files (music, personal and corporate voicemail, etc.)
- Digital photos
- Digital video
- Website metadata (accessibility tags, search engine tags, etc.)
- *Document* metadata (revision histories, document properties and statistics, etc.)
- Internet browser files such as "favorite" URLs, search histories, website cookies, etc.
- Instant message "buddy lists"
- Saved instant or text messages
- Online purchase histories
- Backup tapes, e-mail archives, etc.

While the examples of information stored in a personal or mainframe computer are clearly ESI, keep in mind that ESI is also found in other devices such as cell phones, iPods, BlackBerries, PDAs, etc.

It is interesting to note that while the new eDiscovery rules address how (that is, in what form) ESI should be produced, they don't define or identify what exactly should be retained.

What types of information are stored electronically in *your* company?

ESI Retention and Destruction

A situation occurs in many lawsuits where the plaintiff requests copies of ESI they believe might contain information relevant to the claims in the case, but the defendant claims the ESI has been deleted, overwritten, or otherwise inaccessible. The old discovery rules contained no guidelines about when companies could destroy ESI (such as reusing backup tapes, deleting e-mail archives, etc.), which gave defendants an opportunity to say "Sorry, we deleted that" and plaintiffs an opportunity to ask for all ESI the defendant stored anywhere. This omission in the eDiscovery rules led to accusations of willful destruction of evidence by plaintiffs and complaints of "overly burdensome" ESI production requests by defendants.

While the new eDiscovery rules do not specifically say companies "should" or "must" do anything, they do provide sanctions if your company cannot demonstrate that destroyed or otherwise inaccessible ESI became so in the course of normal business practices (as opposed to the potentially incriminating ESI being purposely destroyed).

The authors believe that if a lawsuit is filed against a company and the plaintiff requests copies of ESI that is no longer accessible, the only way companies can show that ESI was destroyed in the normal course of business is to have formal ESI retention policies that precisely specify what information is saved, the format in which it is saved, where it resides, how long it is maintained, and how and when it can be destroyed.

The authors also advise that in order to ensure compliance with their own ESI retention policies, companies should periodically spot-check or audit affected departments at regular intervals.

Note: While this article has been addressing electronically stored information, companies can also have retention and destruction policies in place for paper-based documents. Check with your company's legal staff about how you might help ensure your company is prepared.

How ESI Is Produced

The new discovery rules also include provisions on how ESI should be produced.

During the discovery process under the old rules, the plaintiff's attorney could ask what data the other side stored electronically then wait to see what (and when, if ever) ESI the defendant would identify. This would often take months—longer if the opposing side was intentionally trying to draw out the process.

For example, the authors were involved in a hit-and-run personal-injury case where the plaintiff suspected that the defendant (a trucking company) had global positioning system (GPS) navigation data in their mainframe that would identify and place one of their trucks at the scene of the accident. Requests for the defendant to identify what GPS and vehicle data they tracked were first refused, with the defendant claiming no such data was saved the past few days. This was in direct conflict from witnesses who swore under oath that they had seen GPS data that had been stored for months, if not years.

When the plaintiff asked the trucking company to produce a copy of their database so the plaintiff could see for themselves, the defendant protested, saying that producing that much data was "unreasonable," "overly burdensome," and that they weren't about to allow the plaintiff on a "fishing trip" through their data. (A judge finally ordered the defendant's computer experts to meet with the plaintiff's experts and to answer the plaintiff's questions about how and where GPS data was stored in their system, but it took *months* and many hours of the plaintiff attorney's time to get just that far.)

Adherence to the new eDiscovery rules could have prevented such tactics. Had the federal rules been in effect at that time, the defendant's attorney would have been required to discuss what ESI was available at the start of the discovery process, and the defendant might have been sanctioned for destroying evidence and the jury (had the case gone to trial) given instructions to construe the missing data as evidence harmful to the defendants.

ESI Production

The new discovery rules also stipulate that ESI must be produced in the format in which it was originally created or in some other form that does not degrade the usefulness of the evidence to the requesting party. In one case with which the authors were involved, a party subpoenaed the other party's e-mails on the matter in question. The other party complied by *printing* every e-mail for each person in question, scanning each e-mail into a bitmap image, and then dropping the scanned image into individual PDF files. (One per page!) It was an act that made the e-mails virtually impossible to search short of opening and *reading* each page (OCR software proved ineffective)—and the company produced *thousands* of e-mails.

These types of obstructionist practices are prohibited under the new rules.

Is your company considering migrating legacy documents to a new authoring tool or environment? If so, be sure to keep copies of the files in their original formats, and keep the tools that can process them as you may well be asked to produce the files in the original format.

Also remember to archive any "keys" or hardware devices, if any, required to run the authoring tools—the original hardware you are using to create backup tapes (tapes are seldom readable by newer models).

These federal rules are already in place. Is your company prepared?

What You Can Do

There are other provisions in the new discovery rules, such as rules dealing with unreasonable discovery requests and for safeguarding a company's trade secrets, but the preceding rule highlights the provisions most critical to businesses.

So what does all this mean to you, the technical communicator? Plenty!

First, you should be aware of the new eDiscovery rules so that you can comply with your company's document-retention polices (if any). Next, your company may not be aware of the new eDiscovery rules, so there is a good chance you may need to write requisite policies or even spearhead the effort for your company to prepare.

What an opportunity to increase your value and sphere of influence in your company!

Steps to Take

Here are steps you might take. Check with your company's legal team to see which are applicable to your situation.

1. Make yourself familiar with the new eDiscovery rules (see "For More Information" at the end of this article).
2. Take a copy of this article to your company's legal counsel and see if anyone in your company is preparing your company to comply with the new rules.
3. Offer to spearhead the effort to make your company prepared.
4. If counsel accepts your offer, explain the situation to your boss and get his/her approval as well.
5. Working with your company's attorney, put together a high-level presentation on the new eDiscovery rules and start briefing management (starting with your boss) and working your way up the organizational chart as far as needed to get "buy-in" from stakeholders. The presentation should include the following:

    a.   A summary of the new eDiscovery rules and the liabilities that can result from your company not being prepared

    b.   The actions all companies need to take to prepare

    c.   A request for the resources needed to research what it will take to make *your* company prepared

    d.   Note: Do not fall into the trap of giving an estimate off the top of your head. Just as one should create a document plan for a complex documentation project prior to giving a firm estimate, you should research how much ESI your company has before estimating how long it will take to inventory and document it.

    e.   A request to be formally appointed as the project manager for the project

Note: If additional resources are needed from outside the company, contact a technical staffing company for temporary resources, or outsource the whole project to a company that provides eDiscovery support, such as **www.ElectronicEvidenceRetrieval.com**.

6.   Create a plan for the project. (If you are new or expanding into project management, ask a senior project manager to mentor you.)

7.   Issue (or draft for someone else to issue) an announcement instructing everyone in the company to cease destroying ESI (as practical) until new ESI—and document-retention policies can be written and issued.

8.   Organize an internal "SWAT team" to identify the areas of the company that contain ESI. The size of the team needed will depend on the size of your company. It may be just you interviewing someone in your IT department, or you may require representatives from various departments/divisions of the company. In either case, be sure to include your company's legal counsel on the team. Keep the legal counsel in the loop as you proceed and ask for clarification about the eDiscovery rules as needed.

9.   Begin inventorying all types of ESI at your company. (See the "For More Information" section for links to sites containing sample retention policy and info-gathering forms.) If available at the time of the inventory, also record how the ESI was created, modified, or destroyed, as well as any extant policies governing its destruction.

10. Compile the various sublists into a master inventory document, and have the SWAT team review them for completeness.

11. For any type of ESI identified that does not have extant retention and destruction policy, have someone in the department/division who "owns" the data determine what the retention and destruction policies for that ESI should be.

Note: Be aware that it is possible (probable!) that there will be many stakeholders involved at this point, each with a different opinion as to which ESI should be retained and for how long. Consider this as another opportunity to expand your sphere of influence and to practice your conflict management and workplace negotiation skills!

12. Have your corporate attorney approve the format you will be using for the final document and ESI retention and destruction policies.
13. Also have your corporate attorney identify what other information you should be collecting. For example, the new eDiscovery rules state that company may not have to produce ESI if the production would be "too burdensome." However, you should be prepared to present a precompiled list of ESI "too burdensome to produce" at the scheduling conference. Consult your company's attorney to determine what would be considered unreasonable for your company, what other information you should *or should not* record, etc.
14. Keep accurate records of your contributions as you proceed through this effort. You'll want written records to show how your company became prepared, and you'll *certainly* want management to appreciate what a monumental task you pulled off when time comes for your annual review and pay raise!
15. If possible, enlist one or more additional technical communicators to draft or review the policy and procedures to ensure they are well written, understandable, measurable, and maintainable.
16. Once you have completed a draft of the ESI inventory and corresponding retention and destruction policies and procedures (as policies for paper-based documents if your company needs them as well), gather as many of your company's legal team as applicable (both internal and external counsel) and brief them on how you've prepared.
17. Once the ESI inventory and retention/destruction policies have been written, create a policy and procedure for auditing them on a regular basis. (Again, consult your company attorney for what would be considered "regular" for your situation.)
18. Get an appropriate (senior) company official to approve and sign the ESI inventory, retention/destruction, and audit policies and procedures.
19. Once approved, issue formal policies and procedure for each specific area of your company, as well as those that apply to everyone company-wide (such as when it's OK to archive and delete e-mails and voice mails).
20. Coordinate with your human resources department so the policies and procedures are incorporated into new-employee orientations and trainings.

21. Ensure procedures are in place to maintain the ESI inventory documents and to audit them as required.

22. Conduct a postmortem on how the project went, turn the maintenance phase over to another member of the team (perhaps an assistant you've been mentoring for just that reason), and then look for another high-visibility project in the company with which to assist.

23. Oh, and remember to ask for a raise come annual review time!

FOR MORE INFORMATION

The new eDiscovery rules are available at no cost at http://www.law.cornell.edu/rules/frcp/.

A copy of the book *The New E-Discovery Rules*, which includes excerpts from the September 2005 report of the Committee on Rules and Practice and Procedure and the May 5, 2005, Report of the Civil Rules Advisory Committee can be obtained for $15.00 at www.legalpub.com.

(Jack Molisani's *Intercom* article on metadata in MS Word. "Ed, can you drop this reference in? I'm on the road and don't have it with me.")

Visit **www.ElectronicEvidenceRetrieval.com** for additional articles and resources on the topics of the new rules and computer forensics, including several sample document retention policies.

Chapter 4

Manufacturing-Business Requirement for Entrepreneurs

The most important aspect for making a project a success is to provide clear requirements. In order to accomplish this goal, the team must go out and complete a survey that encompasses system end users and managers. There are several ways to conduct this survey, but our project will be concerned with rapid application development (RAD). RAD is a group-based development plan with four phases: requirement planning, user design, construction, and cutover.

The project team will be composed of a project leader, system engineer, two technical subject matter experts (SME), four system technicians, and the company information technology manager.

Requirement Planning

Before the beginning of the project, there will be a kickoff meeting to discuss the objects and the schedule of the project. One of the objects will be the fact-finding phase, in which the two SMEs will conduct a survey on the new Oracle 10g. While this task is being done by the SMEs, the system engineer will consult with Oracle and write a procedure in backup of data, migration of data, and installation of Oracle 10g. In the meantime, the four technicians will procure and design the necessary hardware used for the Oracle 10g system. All the processes within this requirement planning are not "set in stone," and any changes will need documentation and approval by the company managers and project leader. Upon the completion of the "kickoff" meeting, the technical team will meet to discuss the design and development of the Oracle 10g system installation.

User Design

The system analyst and project engineer will use CASE tools to achieve and complete their design phase. They will consult with end users to gather more facts about how the system will be used and manipulated. With the newly installed Oracle 10g, the system will be combined into one and provide data access using web portal to internal customers as well as external customers.

Construction

The system technician will procure and install the new hardware for the new Oracle 10g. Any necessary component will be acquired by submitting a request to purchase. This is the first stage of the procurement of the hardware. Only when the approval of this document is accepted can the purchase order be drafted and the hardware ordered. When the parts arrive, the technician will assemble the hardware and install the Oracle 10g. After the installation, the analysts and engineer will configure and program the new software.

Cutover

After the initial migration of the old database to the new Oracle 10g, proper care must be taken to protect the integrity of the legacy database and its tables. The backup solution will be implemented at this time to keep the data in case any failures occur. The current legacy system will be kept online while the 10g is being tested to keep end users productive. There will be a duration of ten working days to fix bugs on the new system. At the end of the tenth day, the legacy will be offline, and the new system will take over its place. It's imperative that the legacy system remain intact just in case the new Oracle 10g fails.

Recap

It is important to keep the project on schedule. In order to accomplish this task, everyone involved must stick to the game plan. By providing a clear requirement and accurate schedule, the project will be completed in a timely manner.

Process flowchart of the procedures and policy statements

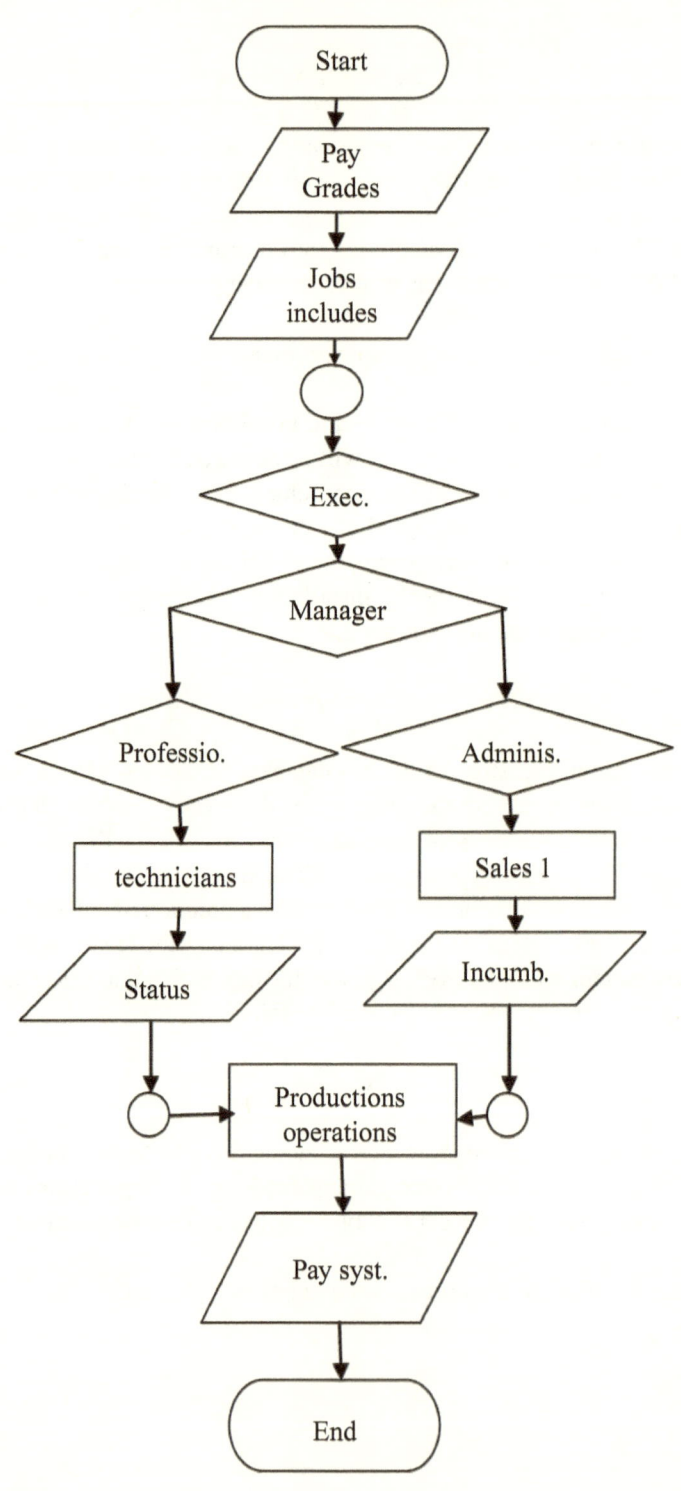

The meaning of some of the words in the flowchart are as follows:

Exec. (Executive)

Professio. (Professional)

Adminis. (Administration)

Incumb. (Number of Incumbents)

Syst. (System)

HUMAN RESOURCE PROCEDURES AND POLICY STATEMENT

An organization's HR system is a constituent of the financial systems package that is designed to keep track of the following employee information: payroll information, W4 forms, date hired, department information, seniority, and all personal contact information.

The organization is oriented with systems development where the following people could be involved: information systems manager, project manager, systems analyst, programmer, business manager, and end users. HR is in charge of making sure the personnel needed are on the job and are trained for it. Each personnel follows the procedures or series of steps of high tech (medical stents, heart valves, etc.) in the manufacturing industry.

The HR department inspects the key elements of performance, interpersonal skills, and standards of ethical practice for business development issues. Training and development records are kept in an Excel worksheet. This part may need to be complimented by an Oracle system training process, which focuses special attention on employee training, where one has a "learning path that identifies the required courses for a desired training goal or certification level. There are recommended learning paths for job roles and one's path based on the Oracle product one wishes to learn." Also, we recommended the requirement that without proper training, no system can be successful because the successful information system requires training for users.

MANAGERS AND IT STAFF MEMBERS

Each recruiter maintains applicant information for open positions. Résumés are filed in a central storage area, and an Excel spreadsheet is used to track the status of applicants.

Workers' compensation is managed by a third-party provider, which keeps its own records. Employee files are kept by individual managers; there is no central employee file area. Managers are also responsible for tracking FMLA (Family Medical Leave Act) absences and any requests for accommodation under the regulation of the public facility, such as any portion of buildings, structures, site improvements, parking lots, or other areas of rescue assistance or evacuation elevators that may be included as part of accessible means of egress on a property. With the company regulations, the compensation manager keeps an Excel spreadsheet with the results of job analyses, salary surveys, and individual compensation decisions.

Employee relations specialists track information about complaints, circumstances regarded as just cause for protest, harassment complaints, etc., in locked files in their offices. The company has its own regulation system based on their philosophy, which is divided in the following manner:

Annual pay adjustments that comply with a process of annual performance evaluations in addition to pay adjustments. This activity occurs yearly; the raises for employees also take affect the first day of their new fiscal year.

Management with a purpose and functionality of each team leader is based on the development activity of regulations, which make sure the complete performance is demanded and must be followed. Every worker and administrator needs a credible performance appraisal system. By using psychometric methods and computer technology, the process of developing such a system can be more efficient and demonstrably successful. For each employee who meets expectations or exceeds expectations, the managers are subject to gain compensations, which are called "a poll of merit increase dollars." This procedure sorts many employees using the chart, where x corresponds to an average percentage of wage (salary) increase. Regulations regarding reward practices are kept to create a team-oriented working environment. This also ensures that people communicate on a regular basis, and this supports and keeps everyone focused on the long-term viability of their organization.

The organization has factors in R & D (research and development), which are critical to the responsibility related to industry team leaders who clarify the industry trends.

There is also a factor of ISO (International Organization for Standardization) 9000 standards, which is very important to be considered in any organization because this is a standard that manufacturing agrees to, where they use an auxiliary to receive *ISO 9000 certification*. This does not exactly mean that the organization's results obtained by performing have a high quality, "only that the company follows well-defined procedures for ensuring quality products."

The key employee relations are as follows:
1. The organization has an "open-door" policy, where workers are free to share their opinions and are even encouraged to share their problems with higher-level supervisors if one was not happy with the management or the supervisor's action.
2. The employee receives a handbook on the first day of hiring that contains the employee policies to follow, such as attendance of organization.
3. Safety experts in the technical field or process are present on location to ensure safety and health to those who need it in the work environment.

Chapter 5

IDP FORENSIC EQUATION AND PRAGMATIC DEFINITION

To work on creating this organization, I realize that life is diverse, and I will describe it with a thought. A leadership strategy, to win, must be based on the following: a vision without practice is a dead dream, and practice without vision is a nightmare. It is fundamental that everything must be worked out in detail to achieve a goal because of the law of unintended consequences. Indeed, in all cases where life and human activity are involved, one basic need is to have a domain or control of any multifaceted activity. Even though the application of intellectual matter is not always dominant to the definition of pragmatism, in this case, it is so. Integrity Data Protection Forensic can be meaningful. The art of influence and directing those actions is done in such a way to obtain one's willing obedience, confidence, respect, and cooperation. These confidences and multitasking are what we call leadership because it is built in a certain way to be acceptable to others.

For many, one might have an interpretation of their focus or be intelligent oriented for performing only one task. This is debatable for me because Jasper syndrome or Asperger's syndrome is a disease reflected in neurobiological disorders. It is considered a severe disease, one thought to be incurable, but IDP Forensic will argue with this. Nothing is absolutely incurable, so Jasper syndrome can be a mild disease that has nothing to do with the exaggeration of a single intelligence-oriented opinion.

This idea leads us to draw the line of this great organization, IDP Forensic, interdisciplinary and an activity to be able to exist, and this is fundamental to the existence of any issue or science.

IDP Forensic's belief, research, practice, and laboratory can find the truth of anything. The dominant error that something is not curable is not also 100 percent, but its ambiguities can be minimized. And so, it is a simple scientific procedure. IDP Forensic considers practice and research from debatable issues that we can demonstrate with the formula:

IDPF=[(total data-data vulnerable)/total data] × 100%

The computation on this formula will never reach infinity; it tends to not be equal to 100 percent. It also means no security is absolute or equal to 100 percent yet because there are many hostilities that compromise our search to be true and absolute. Based on this idea that IDP Forensic is founded to help and safeguard is the following:

IDP Forensic (Computer Forensic Technology) is a safeguard and fundamental for any of life's endeavors.

Chapter 6

SOME COMPUTER TIPS

How to set up your home page:

1. Click on Tool.
2. Go to Internet Options.
3. Click OK.

How to manage the computer system configure date step by step:

Left-click Select
1. Go to Start.
2. Choose Programs.
3. Click on Accessories.
4. Go to System tools. (e.g., date)

How to set up the header of APA style: INSERT:FIELD:page

How to handle basic computer problems:

1. Press Ctrl, Alt, and Delete. All screen of Windows task manager with information
   VPN = Virtual Private Network

How to check computer user:

1. Right-click My Computer.
2. Click Properties.

Let's say that one relies on a very popular virtual database, for example, from www.auntsiter.com, which in our notion will lead one to a strong and stunning information about one's traits. If someone relies on this information to analyze behavior of generations, one might be wrong in the interpretation of the law regarding discrimination because DNA is not subject to interpretation of statutes of behavior of generations as opposed to meticulous information on the real world done in laboratories to find crime.

Files with extension *C will not work for Windows.

Apple (X codes) = you will find the C Program

When you finish writing a C Program code:

1. You must use a compiler.
2. Build second steps

How to use the C Program/compiler:

1. When you have your code already compiled, you apply build to find an error.
2. Run the program where you can see the output.
3. You may find the result of what you want the program to do.

When you find errors in compiling:

1. Compile.
2. Debug the error by checking and rewriting what is needed.
3. Go back to compile A.

If it shows no error line and seems to be fine, return to step 1.
Execute the final step, and you will get the output of the program.

How to print output on C program:

1. Right-click on left side of window of compiler. Click again.
2. Select all and go to the windows and
3. Paste to the windows?

How to select a file and delete it:

1. Press and hold Ctrl key and select each file.
2. Right-click on mouse.
3. Go to Delete then click.

How to make a regular file to print to PDF:

1. Open the work process.
2. Click Print.
3. Go to Print, and select the Adobe Acrobat and Print. This will create a PDF file.

Some IT considerations:

PPPoE Band connection

The **Point-to-Point Protocol over Ethernet** (**PPPoE**) is a network protocol for encapsulating point-to-point protocol (PPP) frames inside Ethernet frames. It is used mainly with DSL services where individual users connect to the DSL modem over an Ethernet cable.

SOME APPLICATIONS WITH THE USE OF APPLE:

1. Apple/Macintosh laptops and hard drive removal. Hard drive removal from a Macintosh laptop can require disassembly of the entire laptop, and any attempt to physically access the drive will void any warranties.
Apple support—http:/apple.com/support
iFixit (laptop guides)—http://www.ifixit.com/
(Guidance Inc. Conference: CEIC2008, Las Vegas)

Some procedures to consider in the use of Mac:

Mac has the operating system Unix (terminal).
To access it, do the following:
1. Go to application.
2. Click Users Utility.
3. Click Terminal, which will load the Unix operating system.

How to troubleshoot the PC (Windows XP) Internet:
Go to the Internet by doing the following:

1. Go to the Control Panel.
2. Select Network Connection.
3. If you find that the status is "Disconnect," then click "Connect" PPPOE.

WINDOWS

How to troubleshoot the Internet connection off-line (PPPOE) Belkin router:

1. Right-click the local area connection.
2. Go to Properties. Double-click and left-click on Protocol.
3. Go to IP and make sure IP is set to obtain IP automatic.

How to print any hard screen:

1. You must press Shift and hold down and press Print Screen key.
2. Minimize it and go to Windows or Microsoft Word. Choose where you want to place it.
3. Paste by holding down Ctrl key and hit V.

How to handle photos, pictures, and picture resolutions:

Select the page: Print Screen
1. Click Start, then Programs.
2. Go to Accessories.
3. View Paint.
4. Click the cursor on the new window.
5. Hit Ctrl + V.

How to delete the message from the screen:

1. Go to the message. Delete.
2. Edit and prepare deleted message.

How to download the paper:

1. Click on Learning Team Charter.
2. Click on Toolkit Essentials.
3. Go to Learning Team Charter.
4. Right-click on the Learning Team Charter.
5. Choose Save Target As (and one can go to storage whenever one wants) After all, you will be able to open the file in Windows.

This procedure could be used to download team evaluation charts also.

How to zoom any type of document:

You must go to View and select 100 percent, which is usually the standard.

How to cut the part you don't need in the pictures:

1. Go to View. Click picture.
2. Go to Crop. Drag it.

This is a very fine procedure to take out what you don't need in a picture file. Paste it.

How to copy and paste a GIF file:

1. Select the GIF file.
2. Right-click on it and select Open With.
3. Scroll down, select Internet Explorer, right-click on Internet with a picture and copy.
4. Go to Microsoft Word and paste it.

It is very simple to obtain the tools. Click on the pictures and select Show Tool picture toolbar.

How to make a computer play the sounds in Windows:

1. Go to Finder.
2. Click on Performance.
3. Choose System performance.
4. Go to Speech recognition.
5. Click on Default voice check (when you drop the cursor voice agrees).

How to find the serial number of an Apple computer:

1. Click on Apple.
2. Click on Version twice. The serial number will show.

To print the budget sheet or any sheet, you must first copy it and save the file in Word.

Procedure to copy from Microsoft Project:

1. Drag the first cell or click the first cell with your mouse (left-click).
2. Press Control key, and at the same time, click the next following cells until the last one. Remember that the control key should be pressed every time you choose a cell. Move the mouse to the right until you drag all the cells that you want.
3. Click Edit. Copy picture.
4. Click GIF and choose the file. Click the selected rows then OK.
5. Go to Microsoft Word, open the document, and choose the position where you want the file.
6. Click Insert picture, then choose the file (find the file you saved before).
7. Click OK.

If the file is too small, just click in the middle and increase it from each side.

How to insert anything to a location:

1. Click Insert and go to picture. Highlight it.
2. From file, click Insert.

Java—an environment for developing cross platform applications. Java is built into and distributed with every copy of Mac OS X.

COCOA—a set of object-oriented application framework that support rapid development of full-featured, high-performance, and high, reliable applications for Mac OS X.

Legacy documents help developers understand legacy technologies.

New for Tiger—Java Tipper includes an enhanced version of Java 1.4.2.

Procedure on how to get layout paper format in landscape (horizontal):

1. Click File.
2. Go to Page Setup.
3. Click on Paper Size.
4. Then go to Orientation.
5. Click Landscape (horizontal).

Note: to switch from one to the other, you must check the box first.

QUESTIONS AND ANSWERS ABOUT THE COBOL PROGRAM

Q & A

Q. How many divisions make up a COBOL program, and what are they called?

A. A COBOL program has four divisions: identification, environment, data, and procedure.

Q. What is the minimum item that the procedure division must contain?

A. At least one paragraph heading.

Q. What is the purpose of the compiler?

A. The compiler checks your program for COBOL language syntax errors, missing or extra items, and basic logic errors. If everything passes the edit, then the compiler creates the basic machine code that is linked using the link edit program or linked to create your actual running program.

Q. Do all programs work correctly the first time?

A. Of course not! According to an old programmer's supposition, any program that compiles without errors the first time must have a bug! The compiler is designed to catch these coding errors and to allow you to fix them.

How to zip a file:

1. Go to whatever it is you want to zip, and right-click WinZip.
2. Select Add to Zip file.
3. Name it and put it wherever you want it to be placed. Make sure the extension says *.zip*.
4. Click on Add.

RED HAT (FREE SOFTWARE)

How to be able to run DOS (disk operating system):

1. Click Run: cmd.
2. Type IP config.
This will show you the IP address.

How to import the program to Excel:

1. First, you go to Create All Tables.
2. Go to Relationships.
3. To import the file, you must first open Excel. Go to Data. Click Select and Import.
4. OK will import your data from Access to Excel.

Computer test:

1. Master drive test—Microsoft System
2. Slave drives test—to store data only

How to handle the install and the full version of the operating system

How to troubleshoot Internet:

1. Go to Internet Explorer.
2. Click Tool.
3. Go to Internet Options.
4. Go to Set Connection.

How to create the page as your default:

1. Go to Tools.
2. Choose Internet Options.
3. From the address types (web pages), click User current.

Bit—short for *binary digit*. The smallest unit of information on a machine. The term was first used in 1946 by John Turkey.
Example
A byte—8 consecutive bits.

Computers are sometimes classified by the number of bits they can house at one time or by the number of bits they use to represent address.

For example, classifying a computer as a 32-bit machine might mean that its data registers are 32 bits wide or that it uses 32 bits to identify each address in memory, whereas larger registers make a computer faster. Using more bits for address enables a machine to support longer programs.

Graphics are also often described by the number of bits.
A 1-bit image is monochrome, an 8-bit image supports 256 colors or grayscales, and a 24—or 32-bit graphic supports true color.

How to use the same of the Unix commands:

Banner—it is a command on a Mac system that will allow the design feature.
Knoppix is an operating system based on Debian designed to be run directly from a CD/DVD (live CD) or a USB flash drive. It is also an important forensic tool.

How to use the shops program:

1. To make or draw the figure, click on the object to get paint. If making a letter, click on the insert, select the text, move it until you get inside the figure, and start typing.

**Some important information about the Unix and Linux resources:
http://www.unix.org/link_list.html**

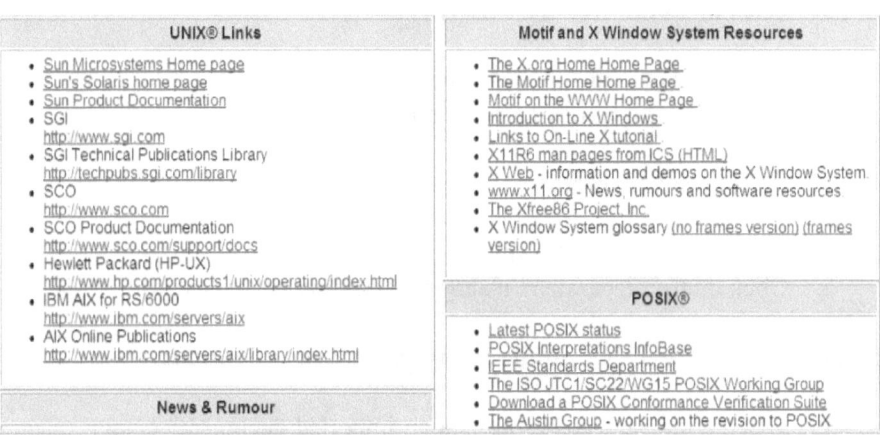

How to find system information:

1. Go to Start (All Programs).
2. Go to Accessories.
3. Go to System Tools.
4. Click on System Information.

How to time your PowerPoint:

1. Go to Slide, then click Transitions.
2. Choose Advance Slide 1 (put the time needed).
3. Click Apply All.

How to handle the Internet connection when all modem connection lights are solid green:

1. Go to Start. Click Run.
2. Type from their IP config.
3. Type Ping, press space, then type Yahoo.com,
   Or type Ping, press space, then type Google.com.
4. You will see the following:

Word: los <0> means you are OK to connect browser but your LAN connection is a problem or your computer LAN settings are not right.

1. Go to Control Panel.
2. Choose Internet Options.
3. Click Connections.
4. Check Never Dial Connection.
5. Click on Local Area Net (LAN).

Make sure all boxes on top are unchecked.
*start run cmd } type: IP Config—/ all

How to find special characters:

1. Go to Insert and click Symbol.
2. Click on the place where it shows a selection of characters.

Some computer forensic considerations about virus symptoms:

1. When a message comes up that an imported document contains macros or a request pops up on whether it should run macros in a document (it is best to disable macros if you cannot verify that they are from a trusted source and that they are free of viruses or worms).
2. A program takes longer than normal to load.
3. The number and length of disk access seem excessive for simple tasks.
4. Common error messages occur regularly.
5. Less memory than occurring is available.
6. Files in a cryptic manner disappear or appear.
7. Strange graphics appear on your computer monitor, or the computer makes strange noises.
8. There is a prominent reduction in disk space.
9. The system cannot recognize the CD-ROM drive although it has worked earlier.
10. The MEM command reveals unusual TSRs loaded into memory.
11. DOS or Windows error messages about the FAT or partition table are displayed.
12. Executable files that once worked no longer work and give unanticipated error messages.
13. Files are frequently corrupted.

Special tools—on how to find errors and replace them:

1. CTRL + will open a box and click the tab where this tool will tell you how many errors you have in the system and you can replace them all.

How to use an Excel sheet:

= Sum (C4:C17) corresponds
When you (+) done need to click enter

To create a new folder, left-click on desktop and then chose New and click Folder to create a new folder. Afterward, you can name this file.

How to create a shortcut:

Click on a highlighted icon and then follow with right click and choose Options shortcut and then you could move it to anywhere on your desktop.

How to add sounds in PowerPoint:

1. Go to Insert.
2. Go to Movies and Sounds.
3. Choose a sound file.
4. Choose where you get sounds from the desktop.
5. Choose the option of how you want it to play.

How to fix the screen of the PC:

1. Right-click on screen and choose Properties.
2. Click Setting and adjust screen resolution. Move it to more (1024 by 768 pixels).

Card scan program:

From the card scan application file □ data exchange □ export all cards to Act! 4.0 very important when you are using card scan 500 machines

Look up example—put in the create date of what you just imported. Leave it at replace lookup and client ok

Edit and replace. For example, <<blank>> in the country field.

Source

How to connect wireless or portable PC:

1. Tech support—click F2 and FN at the same time to be able to access wireless.
2. Connect to network.
3. Click 2wire 081.
4. Enter the security key on bracket found on modem 2wire 8532970732 system key.
5. Please click Connect. For My Wireless, please click on the wireless symbol and enter the password, which is the modem system key.

How to download the sounds from PC to your cell phone:

You must select the mode (sounds) from your cell phone, and from the computer, choose the files and click Transfer.

How to make PowerPoint

1. Go to Instructions.
2. Right-click on the PowerPoint and choose the animation.
3. Then you could maximize the screen slide show.

It is very important to use Excel to make **SUM=B1+B1,** which means you must click on a cell where you want the result to be and apply the formula.

How to create a border:

1. Click Format.
2. Go to Borders and Shading.
3. Choose (box) highlighted.
4. Choose the art you may want and click OK.

ACT DATABASE

How to see the new names updated:

1. Look up by example.
2. In place of created data, you write > and date of the last week.
3. Hit the green exclamation point ~!

How to check if the person is in ACT Database:

1. Go to Contact. Insert note.

How to get AutoShapes:

1. Click Insert.
2. Select Picture.
3. Click AutoShapes.

There is a trick or tip on how to get any company by telephone. Browse this URL: www.gethuman.com.

Note: An acquisition could be done on any crossover system, but it runs very slowly on fast-blocked; in this case, network would show up as "etho." If this is the case, you need to specify an IP address by typing Config etho 10.0.1.

ISDN—integrated service digital network

http://wwww.product activation.

Symatec.com/phone support

To obtain the screen to perform this test, one must turn the power off and restart the PC with the following by typing F12 than now—symptom three service tag click plus to test it, if you found error code (4E63:1813) it means the motherboard is bad.

How to access any PC remotely:

One must have a software program (e.g., Bomgar). There could be others, by downloading a free trial from www.bomgar.com, and learning it by tutorial learning. This software could be used with Windows XP, Vista, and Mac.

Procedure to get it remotely:

1) Other user must connect to Internet.
2) Click your URL. (www.suth.bomgar.com)
4) Click Run.
5) Choose Internet security warning (check on Run).
6) Bomgar C choose 1st.

A recommendation is that one must comment by making sure the timer clock is set at the correct time.

Note: Norton Antivirus and ACT database are owned by Symantec.

VISTA OPERATING SYSTEM

How to use and troubleshoot the sound by using microphones:

One must access his computer. Start and then go through these steps:
1. Click Control Panel.
2. Click Sound.
3. Record and highlight microphone.
4. Click Properties, and then OK.
5. Level and adjust.

Another type of program to use for remote access:

1. Go to www.del.com/remote.
2. Click Agree.
3. Then download.
4. Run.
5. Install.
6. Caller ID can be used to be able to access another person's computer.

Tips for precaution:
Note: Be sure to not download any program or to go to a site that you do not know the purpose of or run on your Internet if you do not have a firewall on your system. There are viruses that replicate themselves, which are called worms. Those viruses that are created by the software are called spyware. This type of virus depends on the user and how he manages his e-mails. Most likely those are pop-ups that can slow down your computer systems. When you click to go to the Internet, it will take one, sometimes two, wrong sites. These can be removed by a technician using tools called remote access.

Security status: Safe! You are protected against most common security threats.
Activex—controls to test your computer for a current antivirus protection.

How to find the type of service providers:

1. Go to Internet.
2. Click Tools.
3. Click Connections.
4. Highlight 192.168.2.84. (This is dial-up only.)
5. Choose OK.

DATABASE

How to replace anything or delete the name in a certain fields:

1. Click Edit.
2. Choose Replace.
3. Go to the field. choose = country = writer blank ok or status ID = status = apply

Work hard but do not stress yourself out.

Special note: When you finish reading this book, give yourself great applause that you got this freedom. Be happy. You can even play (e.g., at Christmastime, go to www.elfyourself.com).

A very important way to find out the last file you opened is by doing the following:

Open the window and click on the file. You will see the last files that were opened.

How to maintain ACT Database:

1. Go to File.
2. Click Administration.
3. Choose Data maintenance.

Windows running on Mac is called Fusion. But the new popularity of the Mac is also partly due to the fact that it can now run Windows along with Apple's superior Mac OS X operating system. That means that if there's a program you need that comes only in a Windows version, you can run it on any current Mac model speedily and with all its features.

There will be a new way to do this. A company called VMware, long the leader in what's called "virtualization"—running multiple operating systems simultaneously on a single computer—will be selling a program called Fusion, which allows Windows and Windows programs to run on a Mac.

Zipping files in ACT

Never have a database open before proceeding:
1. Open ACT file by double-clicking that file.
2. Highlight the file.
3. Right-click and choose WinZip.

4. Go to WinZip and add to zip file.
5. date by finishing with extension = .zip
6. You can save it anywhere.
7. Add, modify, and highlight everything and click Extract.

NOTE: SOME TIPS ON HOW TO CONNECT COMPUTERS

There are various approaches that can be used to connect a Mac in TDM to a Windows system:

Application of TDM is recommended when one is working with Macintosh because it has hard drives that are challenging to remove.

To enable TDM, make sure you boot the Mac and hold down the T key while the system boots. Be patient and keep holding it down until a blue screen with the fire logo appears.

1. Start with both systems off. Connect the systems via a FireWire cable. Boot the Mac to TDM, then boot the Windows system.
2. With the Windows system on, connect the Mac in a powered-off state via the FireWire cable, then boot the Mac to TDM.
3. Boot the Mac to TDM and connect to a powered-on Windows system.

Chapter 7

DEFINITIONS

The capacity of an object is the degree to which that object is transparent. There is nothing more beautiful or nothing I admire more than understanding of phenomenon of transparence of any of life's endeavors. This is something that drew me to the physicist background, *and so you can name this my particle physics that I love so much.* The world is beautiful, and transparency of objects around brings us an understanding and, therefore, happiness. The example below stretches my mind, the same way filters or diachronic filters bring the nonexistent reality of light to look like reality. The way I perceive light, it is such a beautiful thing. In my daily living, I admire and love to see lights, such as the chandelier on my cathedral ceiling in my dining room and the bright against the abysm. I also picture light, and it is directly processed into my mind as an object of integrity, and so is transparency. This is the essence of this book, so it is a *definition* of any subject matter as in this figure: *Diachronic filters.*

MPEG: stands for "motion picture experts group"—an organization that produced a series of standards designed to reduce the storage and transmission requirements of digital video.

Multimedia: any combination of text, video, sound, and graphics.

A completely transparent object has an opacity (or alpha) value of 0 percent. A completely opaque object has an opacity of 100 percent.

PNG file: a graphics file that can be used as an alternative to a GIF graphic file. PNG stands for "portable network group" and supports more graphic features, such as color in images.

Scripts: a programming code that executes one or more commands.

Web server: A web server is a computer that stores, manages, and transfers web pages to other computers (clients) accessing them. Clients can access web servers using a web browser working with media.

TDM: Time division multiplexing is a type of private branch exchange (PBX) used by organizations to switch telephones over physical lines.

GIF file: a graphics file used to transfer images on the Internet whose extensions stand for "graphic interchange format."

QIEM: quality integrity experience management.

JPG or JPEG file: a graphics file whose extension stands for "joint photo export group."

An applet: a small piece of code that can be transported over the Internet and executed on the recipient's machine. The term is especially used to refer to such programs as they are embedded in line as objects in HTML documents on the World Wide Web.

A bitmap: a graphics format that saves information as small dots called pixels. Bitmap can be manipulated one pixel at a time.

Cell: the intersection of a row and column in a table. It is used to display information such as text on a picture.

Managing your site structure is designed to make the file's name management easier. It will tell you, you can rename a file, and the software will change the link names in the applicable files.

Going live: uploading your files once you enter the necessary information into the FTP program. It will connect you to the host computer and send your website files. Once your files are on the host computer, they are available on the World Wide Web.

Search engine:
Fact: Throw a few keywords into your *meta* tags and a top ten spot is as good as yours.

According to *Brandweek* online magazine, search-engine positioning is a fundamental part. It's the baseline. If you are doing nothing else, search engine optimization and key words related to advertising can make up 80 to 90 percent of traffic.

Most of the major search engines now factor link popularity into their relevancy algorithms. As a result, increasing the number of links is still no one's "secret trick" to getting good rankings, but boosting your site's popularity may give it the edge it needs.

In this definition section, I would like to briefly share some remarks to clarify forensically the term economics. Our mission in IDP Forensic is to secure and to use integrity in doing anything, and I also have attention on security that could secure the world economy.

(By stating from Wikipedia, the free encyclopedia)

Economics: The social science that studies the production, distribution, and consumption of goods and services. The term economics comes from the Greek for oiks (house) and nomos (custom or law), hence "rules of the household."

Based on my research, a definition that captures much of modern economics is that of Lionel Robbins in a 1932 essay, "the science which studies human behavior as a relationship between ends and scarce means which have alternative uses." Insecurity means available resources are insufficient to satisfy all wants and needs. If there is an absence of scarcity and alternative uses of available resources, there is no economic problem, and so IDP Forensic is focusing on securing means and values that can make one economically stable.

The subject thus defined involves the study of choices as they are affected by incentives and resources.

Areas of economics may be divided or classified in various ways, including the following:

* Microeconomics and macroeconomics—what exists
* Positive economics ("what is") and normative economics ("what ought to be")
* Mainstream economics and heterodox economics\

The most stable economic regions in the world are in red:

Lack of security leads to disasters, and disasters can be tangent to the economic hardship.

Reference: http://en.wikipedia.org/wiki/Economics

Economic   aftershock: http://w3.newsmax.com/a/aftershockb/video. cfm?PROMO_CODE=CCD5-1

http://tv.breitbart.com/breitbart-news-coverage-of-iowa-caucus/

http://www.youtube.com/watch?NR=1&feature=endscreen&v=3L2513JFJsY

The video to be encoded:

Documentsandsettings\administrator\mydocuments\documentation.avi
1. Go to the Control Panel.
2. Click Next contact.
3. Click on Network connection.
4. Go to Local area network.
5. Click local area network broadband.

To print any slide:

1. Click Print.
2. Check slide.
3. Chose slide #3
4. Print.

Tech support (very important)

1. Check internet connection. Click on Start, then go to Control panel.
2. Go to Network and Internet connections.
3. Go to Network connections.
4. Go to Local Area Networks. It may ask you to set your user name and your password if there were some of these changes.
5. Click Dial.

1. To delete the cookies, click on the Internet page top tool. If there is none, click control T (Ctrl + T). This will bring you up the icons.
2. Go to Internet connection from there. One can manage the following:
    (a) delete cookies
    (b) delete files

IT (INFORMATION TECHNOLOGY)/IS (INFORMATION SYSTEM) MAKES THINGS EASY:

This is the chapter that has many simple commands and instructions on how to manage one's system. It is called IT, which makes it easy because it will help one to use a shortcut command that saves one lots of time. For example, if one does not know how to cut and paste, what should happen is one will have to type something letter by letter. But how about if one takes advantage of using only a mouse, and at a certain point, you did not have the mouse to score and highlight? For example, right-click to select and choose Cut, then paste a page. It will be a little problematic, and you know that your time is money. This is why IT makes it easy in this book.

2.0: Mac response makes it easy: How to master Mac OS X's Hotkeys:

Chapter 8

MS IT (MASTER OF SCIENCE IN INFORMATION TECHNOLOGY) AND IT (INFORMATION TECHNOLOGY) PROJECTS ACCOMPLISHED

OPERATING SYSTEMS

What is the operating system used at work? Why did my organization (ProSpring Inc.) choose the operating system they chose?

Our organization's computers run on Windows XP operating system. The fundamental reason for the use of Windows for our organization is that this operating system could and does give us maximum production. Just in these last few years, we have experienced less frustration than ever since our computers have been upgraded to operating system Windows XP.

Operating systems such as Windows in general are much more reliable for a user because it reduces numerous expenses of the organization. There is a versatility to this operating system that is more powerful than any before. It helps that there are less incidents of the system crashing. I can remember having the OS two years ago, which also helped our organization in the way we were able to perform business. It was an old system but good enough for our business at that time—only a 32-bit operating system, but it was graphical, object-oriented, and able to multitask. We then progressed through the year and eventually upgraded to Windows XP.

We are now able to use different software packages that would sometimes conflict before using XP. Our Windows XP, which is the Professional edition, includes the support of the widest range of software and hardware devices, and it comes also with a feature allowing Windows wireless networks, virus protections, and automatic updates. There is one other wonderful feature about Windows XP where one can undo a bad change, which then returns one's PC to Operating Systems Paper, its previous state. This reduces our added efforts in needing to handle a system that is not compatible and that does not do all that Windows provides.

Windows also has a user-friendly interface called GUI (graphical user interface), which is useful for us to be able to train others and for our own use. With operating system XP, we were able to install our application's softwares: ACT, Eudora, etc., and without Windows as our operating environment to run smoothly, we could not use those hybrid applications. Our use of interface of operating system (Windows XP), indeed enhances ProSpring Inc. to be more functional and have the best technical operation for our business endeavors.

Our resource management hardware has been the best ever—that is, in terms of task management such as managing our tasks and file management (managing data and program files). Our utilities and other functions have also been big improvements with the introduction of Windows XP because of its variety of multitask support services and Windows functions.

We benefit from Windows XP with several computing tasks that are occurring in the form of multiprograms to maintain our daily operations, in which we are able to run our DBMSP (Database Management System of ProSpring Inc).

As personal and business-critical applications become more prevalent on the Internet, network-based applications and services can pose security risks to all information resources. Internet security WAN in many cases has not been given the attention that it deserves. Information is an asset that must be protected. Without network security, information will be flying freely, and a company is highly susceptible to lose its important propriety information (asset).

According to the table Service Request-rm-009 presented below:

Service Request SR-rm-009 Internet Security in the WAN
Organization: use of firewalls, ATM, and switches must be integrated
Manufacturing Location: All plant locations requester
Description of Request: An analysis of internet security for Organization
    Manufacturing WAN
Background of Request: Little documentation exists for the Organization
    Manufacturing WAN.
Expected results impact when completed: An analysis of internet security by
    location in the Organization Manufacturing WAN.

I found in my analysis that Organization Manufacture has a technology weakness in their Information Technology which influences the HR departments.

There is no networking in this department. Each manager has his own database set up on Memorandum on Organization Manufacturing's "SR-rm-009, Internet security in the WAN."

There is no proper central file to analyze in each of the facilities of organization in San Jose (main office), Albany, Pontiac, and Hangzhou (China). The weakness of no-networking technology has intrinsic security dangers in the following areas:

TCP/IP—A communication protocol suite for routed WAN networks was designed as an open standard to facilitate communications. One cannot guard a network against message-modification attacks or protect connections against unauthorized-access attacks.

All three offices use old operating systems such as Windows 95 or 98. However, the use of firewalls and ATM switches must be integrated. Through the use of password protection, authentication, routing protocols, and firewalls, security will be enforced.

Indeed, all configuration weaknesses must be handled by Internet services, which will put high-security data on web servers to be used by enterprise transactions when this type of data (social security numbers, credit card numbers) should be behind a firewall and require user authentication and authorization to access. This would include use of major cryptography services.

WAN network security provides benefits in using a RADIUS server, which gives a client/server architecture, allowing all security information to be located in a single, central database, instead of being scattered around. Therefore, the important features for a Memorandum on Organization Manufacturing's "SR-rm-009, Internet security in the WAN."

WAN includes large modem banks and more than one remote communications server that helps the organization service scalability. A network with the appropriate use of devices will fulfill a need in manufacturing at organization. Organization uses different operating systems. Indeed, there could be a choice to apply RADIUS, which is more practical and manageable for several communication servers located throughout the WAN.

The use of WAN technology requires several security applications selectable as Internet protocol security (IPSec) because it also requires protocol or version 6 (IPv6), which allows us to apply the encryption of data in packets as well as the transport tunnel mode. For this reason, the expected results/impact will be accomplished with the integrating and managing of all the ones used above to obtain successful Internet security in the organization WAN.

Memorandum on Organization Manufacturing's "SR-rm-009, Internet security in the WAN."

OUTSOURCING THE IT/DISASTER RECOVERY FUNCTION

ProSpring Inc. was founded by Jack Molisani in 1996. They advertised themselves by asking, "Do you need an IT programmer, engineer, or IT project manager? Our technical staffing division specializes in recruiting high-tech professionals, both contract and perm. We can provide the right candidate with the right experience at the right price. Do you need a technical writer experienced in your industry? Outsource your IT writing project to our staff technical writers" (www.prospring.net 2006).

What is *outsourcing*?

Per definition, outsourcing is the buying of parts of a product to be assembled elsewhere, as in "purchase cheap foreign parts" rather than manufacturing them at home. However, this definition is more general than outsourcing information technology and is more about letting some expert in the field of disaster recovery be responsive to incidents in your organization, which could be affecting your geographic area. If you are dealing with something outside your country, it is called offshore. Indeed, outsourcing IT/disaster recovery will be handled by an individual who has a domain strategy plan and is trained to reduce or minimize dangerous incidents. IT refers to resources such as computers, software, applications, hardware, and expert personnel as the main elements to control effectiveness of a project. Resource materials, as well as human labor, are the first factors that determine the value of an organization by these things interacting with cost-effective plans for increasing the organization's efficiency, which does determine one of the main factors that leads to a decision in proceeding with outsourcing.

With the use of information technology, the users could plan a way to increase the output or outcome of the plan to become more efficient according to applications. In a pinch, effective increases in an efficiency plan of the combats of computer disasters, for outsourcing the IT/disaster recovery function paper, for example, could be carried out with only a small-time investment and produce an immediate return. One could consider the following strategy: management of disasters is a straightforward organization of documents, plans, and drills of dangerous incidents in your workplace. The objective of outsourcing will be to find the best expert or professional in order to increase the ROI of the company. Managing outsourced work in IT is very popular today by offshoring your business to China or to India rather than having the work done here at home because of the cost of living and commerce in the United States. For example, in South Africa or China, the dollar value is 8 times the value of the yen—that's a reason to offshore work. This economic difference stimulates the business owners to contract IT product support for disaster recovery outside the United States. Nevertheless, the plan of action to pursue physical disaster recovery plans could make it impossible to offshore people.

Risks Associated with Outsourcing IT

Outsourcing IT risks will be convenient where one establishes a communication by creating a wide area network (WAN). In any business, an owner expects to make profits, but there are always risks, especially regarding outsourcing. This is often due to distance. For example, where different cultures are involved, there can be political factors at the destination or geographical location where the product is serviced. In fact, a WAN is needed for offshore outsourcing of departments of organizations. Offshoring can be a negative factor in employment matters of a country due to political instability, which can be a risk for the owner if any changes in political situations result in loss of a contract. Regarding jobs being taken away to foreign countries, in this case, the organization who plans to offshore manufacturing jobs will find it is big business to China's benefit and the owner.

Outsourcing the IT/Disaster Recovery Function

Contrary to popular belief, understanding the impact of offshore outsourcing does not require any formal economics training. Most economists and IT professionals assume the following ideal scenario when thinking about outsourcing. Offshore outsourcing IT may be very risky at times when one confronts social-political, geographic, and cultural situations. Social-political influences are a factor to consider in offshoring. Cultural difference in business ethics may create unexpected results in the understanding of tech support from one country to the next.

The Benefits Associated with an Outsourcing Effort

To be the most practical business nowadays, one must extend one's operations to many international locations where the need to negotiate a contract for services will take place to benefit the company owner because of the future projection of revenue. This practice is the new trend of businesses and applies to low-income countries, who charge low wages for their services. The most professional people are usually consultants who can do the work off-site.

The costs originated in an outsourcing agreement and examples of the dollar impacts expected can be more reasonable than creating a department in one's organization to handle that production. Good business management made according to agreements is intended to cut costs by contracting the most expert consultants that can be found.

Let us create the following scenario:

Before outsourcing the IT/disaster recovery, US workers do tasks A, B, and C, and offshore workers are idle. After outsourcing the IT/disaster recovery, US workers do tasks B, C, and D. Offshore workers do task A and some of B. In this scenario, US workers were doing tasks A, B, and C, and offshore workers were unproductive before the outsourcing occurred. After the outsourcing, the US workers no longer do task A, but have a new task, D. Instead of there being an ideal scenario, the US offshore workers are now doing task A and some of B, so therefore, the US workers remain fully employed but mixed in their tasks. Offshore workers are now fully employed with the tasks of A and B. This is what one hopes will happen.

There are two important problems with this scenario. First, the presumption is that US workers who were previously doing task A will easily be reabsorbed into the workforce by doing tasks such as B, C, and D. This is what economists refer to as the "adjustment" process. Unfortunately, the practical problems with adjustments are substantial. For example, let us assume that task A is computer programming for a certain software required in security, which is increasingly moving offshore, and task C is testing hardware, which is in high demand in the United States. This is realistic to accept (Hire and Hire 2005).

DOLLAR IMPACTS THAT MIGHT BE EXPECTED

One does need the strategy of outsourcing and trade that gives the ROI a boost. On a more fundamental level, because trade was never a problem of world development since the world began, divergent levels of development must trade using offshore means as the new way to do business. Indeed, the gain of outsourcing/offshore is tangible as long as there is a good prediction of potential danger in the negotiations that could occur between the two companies. What one needs to consider along with departmental care/management is those factors to minimize future risk.

Implications to the business organizational structure by using an outsourced IT department is a way the organization could extend their functions or enlarge their departments to operate in the worldwide capacity. The organizational size could develop to the size of a corporation that would need to create their own new IT departments. On the other hand, the security issue would be contingent upon how many resources the company has to make every new area they create safe.

Regarding potential personnel issues that may arise, as usual, in any organization, there are security issues that need to be minimized. Building teamwork is needed to have business intelligence. Human resources will have to work very closely with hiring departments and security to make sure criminal elements are thwarted during the hiring process. Telecommunication (network) is an important element of outsourcing/offshore, which needs to be established with security protection, such as firewalls and encryption, if needed. Business organizations must have auditing of a process and review of the use of their systems by personnel in the organization.

CONCLUSIONS:

OUTSOURCING/OFFSHORE IT

Sometimes, in outsourcing/offshore plans, there are cultural and language barriers that can influence the business even though the offshore team may have technical and skillful people to do the work. There will always be a risk in any business to succeed as above, but caution is used to thwart a disaster. To prevent an unfortunate situation, it would be better to review incidents of the past in order to study and to prevent political barriers that could push aside an aspect of good outsourcing/offshore and cause many business owners to not offshore their business. Those who outsource will make more profit by negotiating a business deal with better workers and less amounts to pay those workers, which is considered as cost-effectiveness or big savings.

Regarding outsourcing, here is an example of something that could happen. Years ago, my organization relocated a technical writer to Paris for a company that also exists here in the United States. This new contractor is American and found the cost of living very high in Paris. He then immediately asked to be transferred back to the United States. Indeed, I realize that it depends on geographic, political, and economic situations of a country, and this can play a big role for a company in outsourcing/offshoring to be successful with ROI and happy workers.

BUSINESS REGULATION

Based on collective research that the additional strategy for Alumna Inc. should have been to negotiate a settlement with the family, this would have resulted to a win for the company rather than going to trial. Even though a direct link between the company's violations and the daughter's leukemia would most likely not be established in a trial case, it would have been a better move to work out a settlement with the family than to go forward into a time-consuming and costly trial. Alumna Inc. should have had the PR representative of the company, Diana Richard, along with their legal department, do one of her deft handlings that she was expert at to negotiate that the company would handle all medical expenses of the daughter and to set up a trust fund that would make the life of the family easier with their great burden. The company would also handle their violations to federal regulations.

Instead, Kelly Bates was allowed to bring a million-dollar personal injury lawsuit against Alumna Inc. that could have been handled in a way that was not a big waste of time and money and which caused more of a bad reputation for the company. A jury will often side with a victim against a very rich corporation, so

Alumna Inc. should also be keen as a large corporation in not letting out certain business confidentialities.

The company does also have the evidence of a scientific nature that reports that increased traffic in the heavily industrialized state of Erehwon is poisoning the water of Lake Dira with polycyclic aromatic hydrocarbons. This is a fact and not something that the company is responsible for. It is also reported that the leading American scientific society in its quarterly journal said that the PAH concentrations have been found to be 100 times greater than preurban conditions and pose a danger to animal and human life. Therefore, it should be a practice of Alumna's PR department to show this evidence to the public and to make it known that Alumna Inc. is not responsible for the different allegations made by Kelly Bates, who accused Alumna Inc. of repeatedly contaminating the waters of Lake Dira with a carcinogenic affluent. This was the basis for her lawsuit, that her 10—year-old daughter's leukemia was caused by their negligence.

Therefore, compare those scientific studies with Kelly Bates's allegations that her daughter got leukemia, and you can see there is a strong discrepancy here. There is also another option that should be looked at. It could cost the company more to settle this case because there could be many future suits that could go the same way and require large settlements, so it should be determined how secure the case of the company is before making that decision. An independent arbitrator could review all the information and decide on a different path that may cause less liability for the company.

Enhanced Project Plan with Resources, Cost, and Written Report will be paid, as usual, by organization as they are employees and will be doing a regular basis work.

The following resources: labor and software applications with a cost of $75,000 and $30,000 respectively are the only costs in the previous cost budget approved last week by organization executives.

The $75,000 + $30,000 = $105,000 total cost of project will be paid for the Oracle 10g database as is shown in the "Enhanced Project Plan with Resources, Cost, and Written Report 3" breakdown in the next section.

Identify the resources and proposed costs needed to complete the project.

Team Member:
Mario Nabliba, IT project manager

CONCEPT

This project plan needs to integrate the current organization's HR system with a new system capable of producing the proper business operation. The previous HR system was spread into different computer management PCs that did not allow workable coordination between staff and facilitate the growth of the company. I plan to use resources and HR end users to get one single and useful system set up with new software and computers and move the past routine spreadsheet applications into the new Oracle Database 100. There will be enough resources used to develop any situation that could be encountered by the new application system to demand that any needed trial-and-error is worked out and that normal operation will not be rejected.

There are stakeholders that will be involved with a high degree of interest who will help to get the work done to minimize the cost of the project and ensure that Enhanced Project Plan with Resources, Cost, and Written Report 4 does not go over budget. These are the in-house costs mentioned above. The plan will propose to use approaches such as interactive, overburden, top-down, and bottom-up. While the project is progressing, I will make sure the baseline is stable. This action will lead to ROI (return on investment) being acceptable and keeping the cost-control method feasible. The plan will establish the procurement management of the 10g database from Oracle and the outsource of work, so the system is set up perfectly by Oracle staff and is at the most reasonable cost.

The purpose of assessing the scope of this plan is to impact the analysis of this project integration plan. To accomplish this objective, one needs to have knowledge in the area of quality, which will include human or nonhuman resources. This is everything needed to finish the project, such as cost, time, communication, etc.

As is stated in the previous weeks of the project plan, I will continue to run the project on a specific time frame—an estimated schedule of 24 hours, and each PM will take a shift so the project can be done in approximately three months.

Enhanced Project Plan with Resources, Cost, and Written Report

BUDGET

My expense budget was prepared according to calculated costs of all resources that will be used for the project, and costs will continue to maintain the activities defined in the planning process. We will modify costs during the planning process to reflect changes that always arise. This effort could be helpful by cooperating with the experienced control people in the information of project services budget.

Items such as day rates per person will be established in the budget preparation and will be used in costing the planned activities.

Our budget plan will be verified and consistent with successful projects of the past so nothing is overlooked and we avoid faux pas. The entire cost of the new HR system is estimated at $105,000, which was arrived at by estimating $75,000 for labor and $30,000 for the system software application licenses. My goal is to have the budget so close to perfect that the project actually comes in at 10 percent less than what is estimated. No budget can be exactly precise, but when one follows the baseline plan, there is an emergency control to change that is allowed.

I am counting on the three IT project managers that we plan to use from Oracle who have many years of experience being able to speed up the project to complete in about 75 days rather than 6 months. This is one of the main reasons I wanted to go with the experience of Oracle.

The integration project will continue to follow WBS (work breakdown structure). At this point, the management team has specified, on the Microsoft Project sheet, the tasks and subtasks and laid out the sequence where one task Enhanced Project Plan with Resources, Cost, and Written Report 7 together forms of software development. That will bring HR an accurate system to manage different software applications that may be needed. The new database, 10G, is set up so it will correspond with any new released version of software. The 10G DB is a system that can help a company increase their ROI an estimated half a million dollars per year.

The new Oracle database is the largest storage DB with 1.5 trillion bytes—it has a high-speed transmission rate and can handle as much as double the amount of data as before. This setup will increase the speed of communication and the management of the HR system. It will be reliable and not crash easily, which could be a big change in the cost-effectiveness for any corporation, and at the same time, the trade-off on the price with the triple constraints (money, time, resource) will be savings for years to come.

ORGANIZATIONAL POLICIES

The organization policies contains the greatest leverage for HR in the following ways: the organization plan will provide them with database features where one is able to access the policies critical to supervisor confidentiality, to check the updated handbook, technical safety policies, health work policies, and all will be accessed on the new 10g Oracle system.

This plan will provide HR with a more secure system to accomplish their goals that will share the responsibility of the organization's HR management and the users of the new system, and have better resources. This database also includes the protection of thief encryption, the protection of transaction, and the avoidance of pitfalls.

Enhanced Project Plan with Resources, Cost, and Written Report will not be initiated until the prior task is finished.

PROJECT PLAN DEVELOPMENT

Organization's brief historical information: "The company's HRIS system was formed in 1992. It contains a financial systems package to track and store the following information: personal information (such as name, address, marital status, birth date, etc.) pay rate, personal exemptions for tax purposes, hire date, seniority date (which is sometimes different than the hire date) organizational information (department for budget purposes, manager's name, etc.) vacation hours (for nonexempt employees).

"Employee files are kept by individual managers; there is no central employee file area. Managers are also responsible for tracking FMLA (Family Medical Leave Act), absences and any requests for accommodation under the ADA. The compensation manager keeps an Excel spreadsheet with the results of job analyses, salary surveys and individual compensation decisions" (UOP Virtual Organization).

Our assumption is that we could develop the new organization HR system that would integrate the dispersion of management files from different personal computers to a simple integrated database system. One possibility from Oracle is their technology version, which will help each phase of the development in a series of steps with a result of multiple users being able to use the system. The new system will carry a large-scale storage in a secure and controlled environment with a single source, involving the benefit of use that is incomparable to the old organization HR system. It may be necessary to blend Enhanced Project Plan with Resources, Cost, and Written Report. Staff members will be aware of the daily progress of the project plan. It will accomplish the project life cycle phases (planning, initiation, controlling, execution, and closing).

Documentation—the plan is to provide HR with the automated documentation as follows:

Focus on understanding the use of our software and hardware, writing, and editing source files to the system. Source from 2 existing Server HR and Finance ("2 ea IBM H520 blade SVTS 2XXEON") with only 2.8 GHz and 1 GB RAM will store new files for a new system—Oracle 10g source server. A server side of system automation integration will explain the following features:

1. Generating output
2. Copying files to a review site web server
3. Testing the quality of the source and output files

The new system will have features on how to train the staff members on the system, online training, configuration, design of prototypes, blueprints, and schedules. Our plan for Organization Manufacturing Inc. HR must include the automatic update system online to keep it accurate and efficient for use on a daily basis of operation.

Timeline Planning: for example, starting November 1, 2011, to January 15, 2012

ENHANCED PROJECT PLAN WITH RESOURCES, COST, AND WRITTEN REPORT WITH FIREWALLS

Controlling and monitoring features will track the corrective actions of reporting and monitoring to increase efficiency, which will help management have privacy and security. This will help to maintain their stable costs.

IMPLEMENTATION

Project plan execution/time line—at this point, the management team is to relay the project plan that will be completed. The old HR system is wiped out by the process of formatting the hard drive. The system will be tested completely and a new Oracle 10g DB will be installed.

According to Microsoft Project, this phase of implementation is named "implementation design," and I plan to do this in twenty days. Each of the tasks needs to be completed before the next one is started. This is called finish-to-start. One could describe the implementation effort as follows:

1. Performing structure analysis for the new HR new system database
2. Providing transition support of the old database and validating implementation against success criteria
3. Creating a change management process by identifying implementation goals and metrics, by CASE (Computer Aided Software Engineering), and by other tools to put into HR

DELIVERABLES

Up to this time, the deliverable plan is following the sequence of the tasks, which could maintain the valid use of the different methodologies and techniques. I am planning the milestone report, which will be a demarcation of progress so the Enhanced Project Plan with Resources, Cost, and Written Report.

CLOSEOUT

The system will be finally tested on month/day/year and will go live on month/day/year.

Therefore, the project will be finished in less than three months. Integration change control will be the project plan, which coordinates the alteration of the old system and completes a special revision so that if any critical situation comes up, it will be handled immediately. The new system will complete the closeout phase by recommending that HR continue maintenance of the system (resources) and cost to be effective.

ENHANCED PROJECT PLAN WITH RESOURCES, COST, AND WRITTEN REPORT

Here is a sample project:

UPDATED HR INTEGRATION PROJECT PLAN AND WRITTEN REPORT

Team members
Mario Nabliba, IT Project Manager/Oracle (pretend that I am from Oracle)
Tim Crowley, IT project manager/Oracle
Organization Manufacturing Inc.
Integration Project Plan and Written Report Concept

Organization HR has presented a business case to integrate a new system with the existing HR system per a memo from Hugh McCauley, chief operating officer of Organization Manufacturing Inc. He has demanded a clarified project plan that is to contain "all tasks, resources, schedule, and budget." I decided to define the roles of responsibility for each person. Since all are IT project managers from Oracle, we will proceed with the integration project planning with a kickoff meeting. The tasks will be assigned and agreed upon by the three IT managers. It will be decided that the project will run on a 24-hour schedule with each taking a shift and each task being handed to the next manager so the task is done around the clock.

The budget estimate of $105,000 was arrived at by estimating $75,000 for labor and $30,000 for the system software application licenses. The three IT project managers planned on and sent from Oracle have many years of experience and are able to speed up the project to complete in about seventy-five days rather than six months also due to the round-the-clock schedule.

The integration project will be according to WBS (work breakdown structure).

Updated HR Integration Project Plan and Written Report where the three IT PMPs (Project Manager Professionals) have a clear view of the project scope. This implementation will integrate Organization Manufacturing HR.

Specifically, this refers to integration, which will help HR to have a more reliable database, which includes benefits information, payroll system, pensions, and a system of maintaining the payroll deductions and rates of insurances plans, health plans, employee personnel records, and employee seniority.

PROJECT PLAN DEVELOPMENT

Employee files are kept by individual managers; there is no central employee file area. Our assumption is that we could develop the new organization HR system that would integrate the dispersion of management files from different personal computers to a simple integrated database systems. One possibility from Oracle Updated HR Integration Project Plan and Written Report 4 is their standard 10g DB, which is a system that can help a company increase their ROI and increase the revenue an estimated half a million dollars per year.

I can make this claim due to the following realities:

The new Oracle database is the largest storage DB with 1.5 trillion bytes—also, it has a high-speed transmission rate and, therefore, can handle double the data that could be handled before. This system will speed up communication and the management of HR. Also, the system will be reliable and is guaranteed not to crash easily, which could be the tradeoff on the price with the triple constraints (money, time, resource).

ORGANIZATIONAL POLICIES

The plan is to provide HR with a more secure database, including the protection from thief encryption, the protection of transaction, and avoidance of pitfalls with firewalls. This will help management have privacy and security.

IMPLEMENTATION

Project plan execution/timeline—at this point, the management team is to convey the project plan that will be completed. The new system is wiped out by the process of formatting the hard drive. The system will be tested and a new Oracle 10g DB will be installed.

In this phase of implementation, according to Microsoft Project, we named this phase "implementation design" and plan to do this in twenty days, which would be perfect for the new HR database system.

Updated HR Integration Project Plan and Written Report 5

DELIVERABLES

It could be valid to use the different methodology and techniques, such as milestones, staff meeting, brainstorming and questions/answers (Q & A) board.

DOCUMENTATION

The plan is to provide HR with the automated documentation as follows:

1. Focus on writing and editing source files.
2. Copying source files to a source server.
3. A server side of system automation integration, which will explain the following features:
   a. Generating output
   b. Copying files to a review site web server
   c. Testing the quality of the source and output files

The new system will have features on how to train the staff members on the system, online training, configuration, design of prototypes, blueprints, and schedules. Our plan for HR must include the automatic update system online to keep it accurate and efficient for use on a daily basis of operation.

PROJECT PLAN EXECUTION/TIMELINE

At this point, the management team is to convey the project plan, which will be completed on the following schedule:
Timeline Planning: start November 1 to January 15
Updated HR Integration Project Plan and Written Report

CLOSEOUT

The system will be finally tested on January 15 and will go live on January 16. Therefore, the project will finish in less than six months. Integration change control will be the project plan, which coordinates the alteration of the old system, completes the project, and is recommended to be consistently maintained by the HR department.

Candidates and clients will be able to e-mail from the same system. All information needs to be online so all personnel in all locations can access the same data.

Here is a sample project:

This week ProSpring Inc., in contact with QIEM Corporation, proceeded into the first phase of the SDLC, which would include the request for proposal, department request, and forming a committee to see this project through. By adding last week's determinations into the scope and objectives, we can achieve a proposal and statement of work contract.

ASSUMPTIONS

This project needs to be completed by October 15. The economic feasibility study has been done, which shows the budget for this should not exceed $4,000. The blueprint report with accurate costs, timelines, and responsibilities needs to be laid out and approved by management.

Customization is their second step, where QIEM will provide ProSpring Inc. with usability testing and design refinements to make sure the end user becomes more aware of the functionality of the new system and to get it ready for implementation for productivity and workflow. There will be database changes that need to have quality checks done.

Data import strategy is the third step where QIEM, from their previews, should be able to develop the basic operations and experimentation data integration links to core business systems that need to be set up for the application system at ProSpring Inc. The training plan for ProSpring Inc. will be done by their own staff.

REQUIREMENT-GATHERING PLAN

What does it mean to begin gathering requirements for a system development project? Requirements gathering for system development (SD) is one of the most important phases of SD. If the requirements are not gathered fully and accurately, the rest of the project is a waste of time since it does not produce what is actually needed.

I began to gather what ProSpring Inc. needs as a system that will handle our recruitment of technical personnel for clients. So far what I have found out is the following:

One can send e-mails from within the contact management system. E-mails are recorded in the contact record. The system must have its own e-mail client not using Outlook Express.

If it has it own e-mail client, we need the new system to be able to make new e-mail boxes easily. This system needs to enable one to drag and drop incoming e-mail to the appropriate mailbox. We want to save all e-mails for a company in one mailbox so we can see at a glance the e-mail traffic for that company.

The current system at ProSpring that needs to be replaced uses software from Sonic Recruit, where jobs are posted and where candidates apply and are then submitted to clients. Eudora is used as the e-mail client, where all communication to candidates and clients are kept. ACT is used to send mailings to clients. We need a system that will incorporate all three of these into one system. We would like to have all contact information in one place for both.

REQUIREMENTS-GATHERING PLAN

Technical writers will have the responsibility to document and to detail the training guide to make it much easier for our users. This will be very important for both training and maintenance, and this will also help ProSpring to avoid the expense of the training that QIEM provides to the employees of a company, which is at least one hour per user. One other factor is the hardware compatibility check and detailed scooping, which must be done in this step. Regarding the security guidelines, we will follow their security procedures required about the systems to facilitate good process of work and flow of production.

DIFFICULTIES IN GATHERING THE REQUIREMENTS

This unique system design, which is a vertical application software, is not easy to obtain. It requires large expense and requires a software architect to build it, and this is why there is a delay in the project. If it were a horizontal application system, would it be easier to design? This is one of the difficulties we need to sort out. Most systems use Outlook Express, and we would prefer not to; therefore, this is a difficulty we need to overcome.

I found that in dealing with a company like QIEM, which has customers worldwide, there is a tremendous amount of follow-up needed to make the project happen. Appointments need to be scheduled with people at QIEM and ProSpring so everyone is in coordination with the follow-through on this project. Sometimes, several schedules are difficult to coordinate.

While keeping all of the above in mind, we all need to follow the company policy and regulations regarding privacy and follow all phases of the SDLC adopted. We hope to keep gathering this information and to succeed.

REQUIREMENTS-GATHERING PLAN

This plan will bring more effectiveness into our business. In fact, we are dealing with a need to overcome the difficulties for the first time, where we will continue to research step by step to find the improvements that will help to gather necessary requirements.

UPDATED HR INTEGRATION PROJECT PLAN AND WRITTEN REPORT

Here is a sample project:

1. Project Integration Management
2. Project Scope Management
3. Project Time Management
4. LavaCon2005 Conference Automating Documentation—Peter Lubbers Oracle 9

RFP Document 2
Organizational Overview
ProSpring Inc., 2500 Via Cabrillo Marina, Suite 200—D San Pedro, CA 90731
e-mail: Info@ProSpring.net
USA

The executive members of the board are as follows:

Mr. Jack Molisani, the founder/president and chief executive officer of ProSpring Inc. He is also the founder of LavaCon, an international conference that offers real-world solutions for managing projects. Mr. Molisani also works part-time as a professor at California State University in Fullerton, teaching a class for technical writers and the tools they use, such as FrameMaker and RoboHelp. He has been running ProSpring Inc. since 1995.

Ms. Suzanne Nabliba is the vice president of ProSpring Inc., a professional counselor and financial analyst who owns part of the company. There are positions supporting the areas of personnel, promotion and marketing, treasury, the production area, a division that qualifies and corrects the staff and a division for procuring new clients.

ProSpring Technical Staffing was founded in 1995 to provide technical staffing and technical writing. Our mission is to provide quality documentation and training materials and to have quality candidates recruited and placed. LavaCon, which was founded in 2003, has the mission to provide training to project and documentation managers. Currently, ProSpring has approximately 2,000 clients.

Last week, ProSpring Inc., in contact with QIEM Corporation, proceeded into the planning of SOLe, which included the request for proposal, department request, and forming a committee to see this project through. By adding last week's determinations into the scope and objectives, we can achieve a proposal and statement of work contract.

Our target for this project is our essential skills needed from both sides, in which we selected the project sponsor who brings the multiple groups together. QIEM's senior engineer already signed the consensus agreement of the project. By brainstorming activities, the QIEM system architect agrees with his team and our project manager where they addressed the issues of our existing systems. They found out that our Pentium III PCS, equipped with 24X CDR-Rom and 48 MB of RAM, qualifies us to put in their recommended system. The project documentation has been approved to be used for new application software. This includes a description of real-time system requirements and how they relate to commonly used real-time constraints. This will be able to support the new system design and the constraint programming of our organization.

REQUIREMENTS DELIVERABLES

The use of the training and documentation from QIEM for our users has been negotiated as a main issue. QIEM will provide support, training, and documentation for free after all installation has been handled, and we will receive tech support during the first 30 days for free. The result is very cost-effective for our organization. Upon conclusion of the project, all material developed by the RFP Document 5 project team can be enhanced as a part of the documentation for ProSpring along with the detailed documents from QIEM.

Continuation of sample project.

ASSUMPTIONS AND AGREEMENTS

This project must be completed by October 15, 2012. The budget for this project is already approved. Our supplier (QIEM) bids will not exceed $4,000 although the job, work, and policy will be assured by ProSpring management team. The design and technical issues and the related prototype work are all outsourced by QIEM to provide the needed new system. This new system is to be implemented to test our existing equipment (system) and approved by ProSpring.

Wiping or formatting the hard drive will be the best way so we can keep our existing hardware and operating system (Windows XP) as the main environment to running the new vertical application. Based on the new operating system, it would be more feasible to use a networking system that will allow the five users to communicate to others. In our service center, our tech-support person will be at the same time the system administrator to maintain the daily service and operation of ProSpring.

ProSpring has, as a priority, to make the system easy, powerful, and durable.

This is IT infrastructure or availability in which we are interested as the heart of our business operation regarding scalability and performance. We will gain our greatest infrastructure technology for our e-business. We will continue the process to provide in-house maintenance activities to cut and prevent the total cost owner (TCO). We will consider the preventive maintenance as priority to investigate, to analyze, and to catch any eventual problem to our system.

RFP DOCUMENT

In case of a disaster or system crash, ProSpring tech support will use digital software recovery disk supplier with the original equipment (software package) and provide the immediate actions, such as corrective, adoptive, and preventive maintenance, according to the case. If the disaster was caused by a virus or a hacker, ProSpring has the person trained in computer forensics using the digital evidence recovery and conversion tools to handle the situation. Therefore, we could use the proof shown by this tool as evidence in a court of law.

ADMINISTRATIVE INFORMATION

Mario Nabliba
How to apply confidentiality
Request for references
Clarification: on the meeting at ProSpring, e.g., Monday, September 12, 2012, with the executive members of the board
This project is debriefed, agreed, and signed by the president, Jack Molisani.
Proposal Format
September 17, 2012
Submit proposal to:
Submission
Decision maker—president, ProSpring Inc. San Pedro, CA 90731

RFP Document
Section criteria

This project has the important elements combined, such as design implementation and development, and should be "done in a development environment and then moved to the production environment during the implementation phase. The development and production environments should be identical in structure and version."

FAULT TOLERANCE

What is *fault tolerance*?

Fault tolerance is defined as the recompense and diversified design characteristics that are essential to enabling a network, a computer application, and a distributed processing system to continue functioning. At one point, it was synonymous with a system of dependability and continuous processing. Fault tolerance can be included in ProSpring's disaster recovery plans that would follow up on a daily basis of maintenance to make sure the system networking is up and running. Indeed, the power system consideration is one of the most important of fault tolerance. With this application, we will identify its use of the tools, staffing, strategies, methods, plans, and procedures necessary to recover in case of disaster. The faults are assumed as random events that interfere with a system by modifying its behaviors. Fault tolerance techniques are formalized by a context, which specifies how replicas of the system cooperate to deal with faults. The system design is guaranteed to behave correctly under a given fault hypothesis by proving the observational equivalence between the specification of the system design and of a fault-free system (UOP 2005).

POWER SYSTEM

A function of the system according to management and administration facilities, the uninterruptible power supplies (UPS) is used in each office and connected for backup of the PCs in case of power failure or emergency so it will kick up for continuity of the service to ensure no data is lost. This will help also in the stability of the business in continuing to have long-term customers.

NETWORK FAULT TOLERANCE

Most of our computer systems service a broad user base over a fairly wide area. The links between a computer and the user community are critical to the uninterrupted supply of computer resources to that user base. This is the way to keep our virtual private network (VPN) running 100 percent and to keep it communicating across a network that is not unreliable or down most of the time. "The same planning needs to be put into the creation of a reliable network infrastructure which goes according to the design or fault-tolerant computers."

TRANSACTION LOGGING/JOURNALIZING

Our organization uses transaction, logging, and journalizing for a robust strategy for keeping information secure. This is by use of transaction journals and the development of database redundancy so one can have the system performing on reliability matters. This provides the successful features used on most relational database management systems and other commercial database management systems.

The way that transaction logs work—a special data file that records all record changes to a database since it was open. If used in conjunction with regular, periodic backups, a transaction log can be used to rebuild a database that has been lost or accidentally destroyed. If the system loses a key file or disk, the last backup tape can be reloaded, and the transactions can be replayed from the time that the tape was made. This procedure helps to recreate the database as it was just before the file was lost. Indeed, the process of the transaction logs, often called "before and image journalizing," is to keep track of these multiple updates to a database.

DATABASE SHADOWING OR MIRRORING

Implementing other strategies to safeguard information is to perform mirroring or shadowing of data. The mirroring procedure is the near real-time copying of data between two disks, in which one becomes a copy of the first and everything written to one is mirrored exactly on the other. Normally, our organization would be happy to implement this approach by using a disk controller, which does not require any special software to be implemented. Loss of a single drive will not result in data loss if a disk goes bad. The controller simply reads from the other disk, which also helps to increase the I/O bandwidth of the disk device in our system.

RAID TECHNOLOGIES

The redundant arrays of inexpensive disk (RAID) strategy involves copying, or shadowing, or shadowing data to extra disk drivers. Different strategies, however, evolved with the realization that safe copies could be kept with less disk space and safe copies could be effectively maintained with less expense than 100 percent redundancy. RAID strategies 1 though 5 use different methods of striping, data compression, and parity: while recoverable and safe, this introduces severe performance problems. RAID boxes are now specialized and are expensive because of the quality of the technology used. Therefore, it is always the most secure to use online-data reliability.

NETWORK REDUNDANCY (ROUTERS ETC.)

ProSpring uses a VPN with high-quality Dell systems. This equipment meets the organizational standard requirement to meet the redundancy of implementation of multiple types of technology that prevent the failure of one system from compromising the security of information. ProSpring has a tradition of selecting special brands of computers that have a high level of internal redundancy to provide possible reliability needs to be introduced to minimize the dependencies on a single wire or network device (routers) wherever possible. Multiple paths of communication need to be designed into networks.

BACKUP CPUs

Backup central processing units (CPUs) are designed to distribute information over different areas to minimize the potential loss of the other units. If all system resource disks, network boxes, UPS system, etc., are located in the same place, then a single catastrophic could become quite costly. A fire or other disaster would not result in the loss of all CPUs, disks, and service distribution of resources across multiple locations, which would essentially safeguard at least some of the hardware from loss due to a single catastrophic event.

SYMMETRIC MULTIPROCESSING (PARALLEL PROCESSING)

It would be good for our organization to have a strategy of symmetric multiprocessing; however, we are migrating gradually to achieve the level of using multiple CPU boards within their systems. This has many advantages, one being the addition of more CPU boards with many slots to facilitate or to do many jobs that need to be performed simultaneously.

Our technician has a plan to have reserve PCs for maintenance in case one fails.

Symmetric multiprocessing is a computer architecture that supplies fast performance to create many CPUs accessible to having all individual processes done at the same time. Different asymmetrical processing, any not-being-used processor can be assigned any task, and additional CPUs can be added to improve performance and handle increased loads. A variety of specialized operating systems and hardware arrangements is available to support SMP. Specific applications can benefit from SMP if the code allows multithreading. SMP uses a single operating system and shares common memory and disk input/output resources. Both Unix and Windows NT support simple management protocol (SMP). SMP is another name for SNMP2. SNMP2 is an enhanced version of the simple network management with features required to support larger networks operating at high data-transmission rates. SNMP2 also supports multiple network management workstations organized in a hierarchical fashion (www.webopedia.comiTERM/S/ SMP.html, May 18, 1998) Security).

Security is often critical to safeguarding databases of any organization as information needs to be protected regardless of other fault tolerance plans that have been designed into one's system. Poor security can result in accidents or malicious loss of your data and programs. There are many kinds of security: physical security, account security, network security, and file security. Although security policies are almost always within the preview of a security department, an internal response team can provide invaluable input to the usefulness, awareness, enforceability, and effectiveness of security policies.

For ProSpring Inc., the fault tolerance plan can be a variety of strategies that we can use in our business operations to minimize downtime of a system or to avoid or to circumvent downtime altogether. Unfortunately, the incorporation of these features often requires planning and a proactive anticipation of likely failure modes before they occur. Our system fault tolerance refers to the design and configuration of computer hardware so that downtime is minimized. In this area, vendors have been very creative, and a variety of different strategies has been tried over the years. There are many applications for fault tolerance as is included in the introduction and basic definitions of type of design. A fault monitoring system is an engineering procedure for systematically checking for errors and malfunctions in the software and hardware of a computer or control system. Fault tolerance measures hard drives, which represent the single largest failure point in a network. Application of fault tolerance measures will save you from a failed power supply or memory module. It could also be used to prevent processor failures, for having space memory, and for preparing for faulty power supplies.

It is possible to find applications that need little fault tolerance. Redundancy means a high quality of material involved to increase the tolerance and durability of performance capabilities. However, it would be almost impossible to have no fault tolerance capabilities because the system's fault tolerance design or hardware is dependent on assessing it as a failure or as safe even though the system is a reliable system. The network system ProSpring uses is SISCO technology, which is considered the best design for performance. One would need at least the same fault tolerance, design, quality, and configuration of computer hardware in order to minimize downtime.

To be specific, ProSpring, as a small technical organization, should consider fault tolerance as a system, network, software, and database for the prevention of system failures. There are two different basic strategies that can be applied to deal with this situation. Design systems with built-in redundancy so that critical systems have backups that will continue to function should primary systems fail. Design systems that are more reliable or that fail less often. Indeed, our organization chose the second strategy, for reasons of having a more cost-effective system that has 99 percent reliability. Our computer systems have only 1 percent downtime every year.

ENERGY AND THE ENVIRONMENT

This subject of energy and the environment has been one of the great issues that I make and why I have been very supportive of the good initiatives about a safe environment. This was one of the ideas I wrote to my state senator (California), the greenhouse gases effect, as my wife was asking for conservation of trees around our neighborhood, Monterey Hills. There is an amount of impure air in our city of Los Angeles that she thinks we must do something about. I think what we have to do here is to practice energy efficiency. I want to mention the law and policy; nature, law, and society are very important to follow and enforce if it is necessary to do so. Why? Because if the earth is natural, it tends to be more environmentally friendly to humans.

This ought to be the work of social scientists, our lawmakers, and our government regulators and scientists, where both could benchmark applications of these policies. If we reward scholarship programs that apply energy efficiency, it will help to decrease the use of detrimental energy that causes devastation to our people, (e.g., hurricanes and other phenomena that we call natural calamities). In my opinion, it is not naturally caused by itself; however, by human action, that is caused by an elevated temperature built on emissions of CO_2. It is scientifically proven—"does a gas at standard temperature and pressure and exist in the earth's atmosphere?"

It must be done in a cost-effective way. One can apply axioms to sewage treatment, plants, and other pollution prevention and control systems. This will indeed keep our lives and environment safe.

The scientific field, so-called environmental engineering, should strive for more study and educational scholarship reward programs, emphasizing the importance of natural law that prevents damage to the earth and climate change. If there is damage to the earth and the climate due to a limitation in our scope of knowledge, let's then apply scientific research or use alternative energy with cost-effectiveness.

Chapter 9

GREAT QUOTES

The quotations of some great minds will help you be secure and focused on the meaning of any project.

INTEGRITY

No public man can be just a little crooked.

—Herbert Hoover

The Man who is prepared has his battle half-fought.

—Cervantes

I will prepare, and someday, my change will come.

—Abraham Lincoln

Before everything else, getting ready is the secret of success.

—Henry Ford

The essence of knowledge is living it, to apply it; not having it to confess your ignorance.

—Confucius

Knowledge and human power are synonymous.

—Francis Baron

A perfection will be hard or never achieved if one does not network his idea.

—**Mario Nabliba "Crowley"**

dedicated to:

AMERICAN INSTITUTE IDPF

PERSISTENCE

We are what we repeatedly do.

—Aristotle

MANAGING CONFLICT

There never was a time, in my opinion, some way could not be found to prevent the drawing of the sword.

—General Ulysses S. Grant

HANDLING THIS STRESSFUL ECONOMY

The ultimate measure of a man is not where he stands in moments of comfort and convenience, but where he stands at times of challenge and controversy.

—Mario Nabliba "Crowley"

Chapter 10

How to Behave with Integrity and Respect with Other Nations Is to at Least Learn Their Constitution or the Law of the Land

Special Note: Why I chose the UK Constitution here—we have the official English language in common. Most of our citizens in America already know the US Constitution, and I would like us to understand a neighboring nation for business purposes.

The House of Stuart and the Commonwealth (1603-1714)

James I

James I of England

With the death of Elizabeth in 1603, the Crowns of England and Scotland united under James I. In 1567, when he was just a year old, James' mother Mary was forced to abdicate, and James became King James VI. Despite his mother's Catholicism, James was brought up as a Protestant.

One of James' first acts as King was to conclude English involvment in the Eighty Years' War, also called the Dutch Revolt. Elizabeth had supported the Protestant Dutch rebels, providing one cause for Philip II's attack. In 1604, James signed the Treaty of London, thereby making peace with Spain.

James had significant difficulty with the English Parliamentary structure. As King of Scots, he had not been accustomed to criticism from the Parliament. James firmly believed in the *Divine Right of Kings*—the right of Kings to rule that supposedly came from God—so he did not easily react to critics in Parliament. Under English law, however, it was impossible for the King to levy taxes without Parliament's consent, so he had to tolerate Parliament for some time.

King James died in 1625 and was succeeded by his son Charles.

CHARLES I

King Charles ruled at a time when Europe was moving toward domination by absolute monarchs. The French ruler, Louis XIV, epitomised this absolutism. Charles, sharing his father's belief in the Divine Right of Kings, also moved toward absolutist policies.

Charles conflicted with Parliament over the issue of the Huguenots, French Protestants. Louis XIV had begun a persecution of the Huguenots; Charles sent an expedition to La Rochelle to provide aid to the Protestant residents. The effort, however, was disastrous, prompting Parliament to further criticise him. In 1628, the House of Commons issued the Petition of Right, which demanded that Charles cease his use of arbitrary power. Charles had persecuted individuals using the Court of the Star Chamber, a secret court that could impose any penalty, even torture, except for death. Charles had also imprisoned individuals without a trial and denied them the right to the writ of *habeas corpus*. The Petition of Right, however, was not successful; in 1629, Charles dissolved Parliament. He ruled alone for the next eleven years, which is sometimes referred to as the *eleven years of tyranny* or *personal rule*. Since Parliamentary approval was required to impose taxes, Charles had grave difficulty in keeping the government functional. Charles imposed several taxes himself; these were widely seen as unlawful.

During these eleven years, Charles began instituting religious reforms in Scotland, moving it towards the English model. He attempted to impose the Anglican Prayer Book on Scottish churches, leading to riots and violence. In 1638, the General Assembly of the Church of Scotland abolished the office of bishop and established Presbyterianism (an ecclesiastic system without clerical officers such as bishops and archbishops). Charles sent his armies to Scotland, but was quickly forced to end the conflict, known as the First Bishops' War, because of a lack of funding. Charles granted Scotland certain parliamentary and ecclesiastic freedoms in 1639.

In 1640, Charles finally called a Parliament to authorise additional taxation. Since the Parliament was dissolved within weeks of its summoning, it was known as the *Short Parliament*. Charles then sent a new military expedition to Scotland to fight the Second Bishops' War. Again, the Royal forces were defeated. Charles then summoned Parliament again, this Parliament becoming known as the *Long Parliament*, in order to raise funds for making reparations to the Scots.

Tension between Charles and Parliament increased dramatically. Charles agreed to abolish the hated Star Chamber, but he refused to give up control of the army. In 1641, Charles entered the House of Commons with armed guards in order to arrest his Parliamentary enemies. They had already fled, however, and Parliament took the breach of their premises very seriously. (Since Charles, no English monarch has sought to set foot in the House of Commons.)

The unsafe monarch moved the Royal court to Oxford. Royal forces controlled north and west England, while Parliament controlled south and east England. A Civil War broke out, but was indecisive until 1644, when Parliamentary forces clearly gained the upper hand. In 1646, Charles was forced to escape to Scotland, but the Scottish army delivered him to Parliament in 1647. Charles was then imprisoned. Charles negotiated with the Scottish army, declaring that if it restored him to power, he would implement the Scottish Presbyterian ecclesiastic model in England. In 1648, the Scots invaded England, but were defeated.

The House of Commons began to pass laws without the consent of either the Sovereign or the House of Lords, but many MPs still wished to come to terms with the king. Members of the army, however, felt that Charles had gone too far by sideing with the Scots against England and were determined to have him brought to trial. In December 1648 an army regiment, Colonel Pride's, used force to bar entry into the House of Commons, only allowing MPs who would support the army to remain. These MPs, the *Rump Parliament*, established a commission of 135 to try Charles for treason. Charles, an ardent believer in the Divine Right of Kings, refused to accept the jurisdiction of any court over him. Therefore, he was by default considered guilty of high treason and was executed on January 30, 1649.

OLIVER AND RICHARD CROMWELL

At first, Oliver Cromwell ruled along with the republican Parliament, the state being known as the *Commonwealth of England*. After Charles' execution, however, Parliament became disunited. In 1653, he suspended Parliament, and as Charles had done earlier, began several years of rule as a dictator. Later, Parliament was recalled, and in 1657 offered to make Cromwell the King. Since he faced opposition from his own senior military officers, Cromwell declined. Instead, he was made a *Lord Protector*, even being installed on the former King's throne. He was a King in all but name.

Cromwell died in 1658 and was succeeded by his son Richard, an extremely poor politician. Richard Cromwell was not interested in his position and abdicated quickly. The Protectorate was ended and the Commonwealth restored. Anarchy was the result. Quickly, Parliament chose to reestablish the monarchy by inviting Charles I's son to take the throne as Charles II.

CHARLES II

During the rule of Oliver Cromwell, Charles II remained King in Scotland. After an unsuccessful challenge to Cromwell's rule, Charles escaped to Europe. In 1660, when England was in anarchy, Charles issued the Declaration of Breda, outlining his conditions for returning to the Throne. The Long Parliament, which had been convened in 1640, finally dissolved itself. A new Parliament, called the *Convention Parliament*, was elected; it was far more favourable to the Royalty than the Long Parliament. In May 1660, the Convention Parliament that Charles had been the lawful King of England since the death of his father in 1649. Charles soon arrived in London and was restored to actual power. Charles granted a general pardon to most of Cromwell's supporters. Those who had directly participated in his father's execution, however, were either executed or imprisoned for life. Cromwell himself suffered a posthumous execution: his body was exhumed, hung, drawn and quartered, his head cut off and displayed from a pole and the remainder of his body thrown into a common pit. The posthumous execution took place on the anniversary of Charles I's death.

Charles also dissolved the Convention Parliament. The next Parliament, called the *Cavalier Parliament* was soon elected. The Cavalier Parliament lasted for seventeen years without an election before being dissolved. During its long tenure, the Cavalier Parliament enacted several important laws, including many that suppressed religious dissent. The Act of Uniformity required the use of the Church of England's *Book of Common Prayer* in all Church services. The Conventicle Act prohibited religious assemblies of more than five members except under the Church of England. The Five Mile Act banned non-members of the Church of England from living in towns with a Royal Charter, instead forcing them into the country. In 1672, Charles mitigated these laws with the Royal Declaration of Indulgence, which provided for religious toleration. Parliament, however, suspected him of Catholicism and forced him to withdraw the Declaration. In 1673, Parliament passed the Test Act, which required civil servants to swear an oath against Catholicism.

Parliament's suspicions did turn out to be accurate. As Charles II lay dying in 1685, he converted to Catholicism. Charles did not have a single legitimate child, though he did have, while living in Europe, several illegitimate ones (over 300 by some estimates). He was succeeded, therefore, by his younger brother James, an open Catholic.

JAMES II

James II (James VII in Scotland) was an extremely controversial monarch due to his Catholicism. Soon after he took power, a Protestant illegitimate son of

Charles II, James Scott, Duke of Monmouth, proclaimed himself King. James II defeated him within a few days and had him executed.

James made himself highly unpopular by appointing Catholic officials, especially in Ireland. Later, he established a standing army in peacetime, alarming many Protestants. Rebellion, however, did not occur because people trusted James' daughter Mary, a Protestant. In 1688, however, James produced a son, who was brought up Catholic. Since Mary's place in the line of succession was lowered, and a Catholic Dynasty in England seemed ineveitable, the "Immortal Seven"—the Duke of Devonshire, the Earl of Danby, the Earl of Shrewsbury, the Viscount Lumley, the Bishop of London, Edward Russell and Henry Sidney—conspired to replace James and his son with Mary and her Dutch husband William of Orange. In 1688, William and Mary invaded England and James fled the country. The revolution was hailed as the *Glorious Revolution* or the *Bloodless Revolution*. Though the latter term was innacurate, the revolution was not as violent as the War of the Roses or the English Civil War.

WILLIAM AND MARY

Parliament wished then to make Mary the sole Queen. She, however, refused and demanded that she be made co-Sovereign with her husband. In 1689, the Parliament of England declared in the English Bill of Rights, one of the most significant constitutional documents in British history, that James' flight constituted an abdication of the throne and that the throne should go jointly to William (William III) and Mary (Mary II). The Bill of Rights also required that the Sovereign cannot deny certain rights, such as freedom of speech in Parliament, freedom from taxation without Parliament's consent and freedom from cruel and unusual punishment. In Scotland, the Estates General passed a similar Act, called the Claim of Right, which also made William and Mary joint rulers. In Ireland, power had to be won in battle. In 1690, the English won the Battle of the Boyne, thereby establishing William and Mary's rule over the entire British Isles.

For the early part of the reign, Mary administered the Government while William controlled the military. Unpopularly, William appointed people from his native Holland as officers in the English army and Royal Navy. Furthermore, he used English military resources to protect the Netherlands. In 1694, after the death of Queen Mary from smallpox, William continued to rule as the sole Sovereign.

Since William and Mary did not have children, William's heir was Anne, who had seventeen pregnancies, most of which ended in stillbirth. In 1700, Anne's last surviving child, William, died at the age of eleven. Parliament was faced with a succession crisis, because after Anne, many in the line of succession were Catholic. Therefore, in 1701, the Act of Settlement was passed, allowing Sophia, Electress and Duchess Dowager of Hanover (a German state), and her Protestant heirs, to succeed if Anne had no further children. Sophia's claim stemmed from her great-grandfather, James I. Several lines that were more senior to Sophia's were bypassed under the act. Some of these had questionable legitimacy, while others were Catholic. The Act of Settlement also banned non-Protestants and those who married Catholics from the throne.

In 1702, William died, and his sister-in-law Anne became Queen.

Even following the passage of the Act of Settlement, Protestant succession to the throne was insecure in Scotland. In 1703, the Scottish Parliament, the Estates, passed a bill that required that, if Anne died without children, the Estates could appoint any Protestant descendant of Scottish monarchs as the King. The individual appointed could not be the same person who would, under the Act of Settlement, succeed to the English crown unless several economic conditions were met. The Queen's Commissioner refused Royal Assent on her behalf. The Scottish Estates then threatened to withdraw Scottish troops from the Queen's armies, which were then engaged in the War of the Spanish Succession in Europe and Queen Anne's War in North America. The Estates also threatened to refuse to levy taxes, so Anne relented and agreed to grant Royal Assent to the bill, which became the Act of Security.

The English Parliament feared the separation of the Crowns which had been united since the death of Elizabeth I. They therefore attempted to coerce Scotland, passing the Alien Act in 1705. The Alien Act provided for cutting off trade between England and Scotland. Scotland was already suffering from the failure of the Darién Scheme, a disastrous and expensive attempt to establish Scottish colonies in America. Scotland quickly began to negotiate union with England. In 1707, the Act of Union was passed, despite mass protest in Scotland, by Parliament and the Scottish Estates. The Act combined England and Scotland into one Kingdom of Great Britain, terminated the Parliament and Estates, and replaced them with one Parliament of Great Britain. Scotland was entitled to elect a certain number of members of the House of Commons. Furthermore, it was permitted to send sixteen of its peers to sit along with all English peers in the House of Lords. The Act guaranteed Scotland the right to retain its distinct legal system. The Church of Scotland was also guaranteed independence from political interference. Ireland remained a separate country, though still governed by the British Sovereign.

Anne is often remembered as the last British monarch to deny Royal Assent to a bill, which she did in 1707 to a militia bill. Due to her poor health, made worse by her failed pregnancies, her government was run through her ministers. She died in 1714, to be succeeded by George, Elector of Hanover, whose mother Sophia had died a few weeks earlier.

Chapter 11

Direct Quotes Regarding Copyrights

The purpose of this License is to make a manual, textbook, or other functional and useful document "free" in the sense of freedom: to assure everyone the effective freedom to copy and redistribute it, with or without modifying it, either commercially or no commercially. Secondarily, this License preserves for the author and publisher a way to get credit for their work, while not being considered responsible for modifications made by others.

This License is a kind of "copyleft," which means that derivative works of the document must themselves be free in the same sense. It complements the GNU General Public License, which is a copyleft license designed for free software. We have designed this License in order to use it for manuals for free software, because free software needs free documentation: a free program should come with manuals providing the same freedoms that the software does. But this License is not limited to software manuals; it can be used for any textual work, regardless of subject matter or whether it is published as a printed book. We recommend this License principally for works whose purpose is instruction or reference.

1. Applicability and Definitions

This License applies to any manual or other work, in any medium, that contains a notice placed by the copyright holder saying it can be distributed under the terms of this License. Such a notice grants a world-wide, royalty-free license, unlimited in duration, to use that work under the conditions stated herein. The "Document," below, refers to any such manual or work. Any member of the public is a licensee, and is addressed as "you." You accept the license if you copy, modify or distribute the work in a way requiring permission under copyright law.

A "Modified Version" of the Document means any work containing the Document or a portion of it, either copied verbatim, or with modifications and/or translated into another language.

A "Secondary Section" is a named appendix or a front-matter section of the Document that deals exclusively with the relationship of the publishers or authors of the Document to the Document's overall subject (or to related matters) and contains nothing that could fall directly within that overall subject. (Thus, if the Document is in part a textbook of mathematics, a Secondary Section may not explain any mathematics.) The relationship could be a matter of historical connection with the subject or with related matters, or of legal, commercial, philosophical, ethical or political position regarding them.

The "Invariant Sections" are certain Secondary Sections whose titles are designated, as being those of Invariant Sections, in the notice that says that the Document is released under this License. If a section does not fit the above definition of Secondary then it is not allowed to be designated as Invariant. The Document may contain zero Invariant Sections. If the Document does not identify any Invariant Sections then there are none.

The "Cover Texts" are certain short passages of text that are listed, as Front-Cover Texts or Back-Cover Texts, in the notice that says that the Document is released under this License. A Front-Cover Text may be at most 5 words, and a Back-Cover Text may be at most 25 words.

A "Transparent" copy of the Document means a machine-readable copy, represented in a format whose specification is available to the general public, that is suitable for revising the document straightforwardly with generic text editors or (for images composed of pixels) generic paint programs or (for drawings) some widely available drawing editor, and that is suitable for input to text formatters or for automatic translation to a variety of formats suitable for input to text formatters. A copy made in an otherwise Transparent file format whose markup, or absence of markup, has been arranged to thwart or discourage subsequent modification by readers is not Transparent. An image format is not Transparent if used for any substantial amount of text. A copy that is not "Transparent" is called "Opaque."

Examples of suitable formats for Transparent copies include plain ASCII without markup, Texinfo input format, LaTeX input format, SGML or XML using a publicly available DTD, and standard-conforming simple HTML, PostScript or PDF designed for human modification. Examples of transparent image formats include PNG, XCF and JPG. Opaque formats include proprietary formats that can be read and edited only by proprietary word processors, SGML or XML for which the DTD and/or processing tools are not generally available, and the machine-generated HTML, PostScript or PDF produced by some word processors for output purposes only.

The "Title Page" means, for a printed book, the title page itself, plus such following pages as are needed to hold, legibly, the material this License requires to appear in the title page. For works in formats which do not have any title page as such, "Title Page" means the text near the most prominent appearance of the work's title, preceding the beginning of the body of the text.

A section "Entitled XYZ" means a named subunit of the Document whose title either is precisely XYZ or contains XYZ in parentheses following text that translates XYZ in another language. (Here XYZ stands for a specific section name mentioned below, such as "Acknowledgements," "Dedications," "Endorsements," or "History.") To "Preserve the Title" of such a section when you modify the Document means that it remains a section "Entitled XYZ" according to this definition.

The Document may include Warranty Disclaimers next to the notice which states that this License applies to the Document. These Warranty Disclaimers are considered to be included by reference in this License, but only as regards disclaiming warranties: any other implication that these Warranty Disclaimers may have is void and has no effect on the meaning of this License.

2. VERBATIM COPYING

You may copy and distribute the Document in any medium, either commercially or no commercially, provided that this License, the copyright notices, and the license notice saying this License applies to the Document are reproduced in all copies, and that you add no other conditions whatsoever to those of this License. You may not use technical measures to obstruct or control the reading or further copying of the copies you make or distribute. However, you may accept compensation in exchange for copies. If you distribute a large enough number of copies, you must also follow the conditions in section.

You may also lend copies, under the same conditions stated above, and you may publicly display copies.

3. COPYING IN QUANTITY

If you publish printed copies (or copies in media that commonly have printed covers) of the Document, numbering more than 100, and the Document's license notice requires Cover Texts, you must enclose the copies in covers that carry, clearly and legibly, all these Cover Texts:

Front-Cover Texts on the front cover, and Back-Cover Texts on the back cover. Both covers must also clearly and legibly identify you as the publisher of these copies. The front cover must present the full title with all words of the title equally prominent and visible. You may add other material on the covers in addition. Copying with changes limited to the covers, as long as they preserve the title of the Document and satisfy these conditions, can be treated as verbatim copying in other respects.

If the required texts for either cover are too voluminous to fit legibly, you should put the first ones listed (as many as fit reasonably) on the actual cover, and continue the rest onto adjacent pages. If you publish or distribute Opaque copies of the Document numbering more than 100, you must either include a machine-readable Transparent copy along with each Opaque copy, or state in or with each Opaque copy a computer-network location from which the general network-using public has access to download using public-standard network protocols a complete Transparent copy of the Document, free of added material. If you use the latter option, you must take reasonably prudent steps, when you begin distribution of Opaque copies in quantity, to ensure that this Transparent copy will remain thus accessible at the stated location until at least one year after the last time you distribute an Opaque copy (directly or through your agents or retailers) of that edition to the public.

It is requested, but not required, that you contact the authors of the Document well before redistributing any large number of copies, to give them a chance to provide you with an updated version of the Document.

4. MODIFICATIONS

You may copy and distribute a Modified Version of the Document under the conditions of sections 2 and 3 above, provided that you release the Modified Version under precisely this License, with the Modified Version filling the role of the Document, thus licensing distribution and modification of the Modified Version to whoever possesses a copy of it. In addition, you must do these things in the Modified Version:

A. Use in the Title Page (and on the covers, if any) a title distinct from that of the Document, and from those of previous versions (which should, if there were any, be listed in the History section of the Document). You may use the same title as a previous version if the original publisher of that version gives permission.

B. List on the Title Page, as authors, one or more persons or entities responsible for authorship of the modifications in the Modified Version, together with at least five of the principal authors of the Document (all of its principal authors, if it has fewer than five), unless they release you from this requirement.

C. State on the Title page the name of the publisher of the Modified Version, as the publisher.

D. Preserve all the copyright notices of the Document.

E. Add an appropriate copyright notice for your modifications adjacent to the other copyright notices.

F. Include, immediately after the copyright notices, a license notice giving the public permission to use the Modified Version under the terms of this License, in the form shown in the Addendum below.

G. Preserve in that license notice the full lists of Invariant Sections and required Cover Texts given in the Document's license notice.

H. Include an unaltered copy of this License.

I. Preserve the section Entitled "History," Preserve its Title, and add to it an item stating at least the title, year, new authors, and publisher of the Modified Version as given on the Title Page. If there is no section Entitled "History" in the Document, create one stating the title, year, authors, and publisher of the Document as given on its Title Page, then add an item describing the Modified Version as stated in the previous sentence.

J. Preserve the network location, if any, given in the Document for public access to a Transparent copy of the Document, and likewise the network locations given in the Document for previous versions it was based on. These may be placed in the "History" section. You may omit a network location for a work that was published at least four years before the Document itself, or if the original publisher of the version it refers to gives permission.

K. For any section Entitled "Acknowledgements" or "Dedications," Preserve the Title of the section, and preserve in the section all the substance and tone of each of the contributor acknowledgements and/or dedications given therein.

L. Preserve all the Invariant Sections of the Document, unaltered in their text and in their titles. Section numbers or the equivalent are not considered part of the section titles.

M. Delete any section Entitled "Endorsements." Such a section may not be included in the Modified Version.

N. Do not retile any existing section to be Entitled "Endorsements" or to conflict in title with any Invariant Section.

O. Preserve any Warranty Disclaimers.

If the Modified Version includes new front-matter sections or appendices that qualify as Secondary Sections and contain no material copied from the Document, you may at your option designate some or all of these sections as invariant. To do this, add their titles to the list of Invariant actions in the Modified Version's license notice. These titles must be distinct from any other section titles.

You may add a section Entitled "Endorsements," provided it contains nothing but endorsements of your Modified Version by various parties—for example, statements of peer review or that the text has been approved by an organization as the authoritative definition of a standard. You may add a passage of up to five words as a Front-Cover Text, and a passage of up to 25 words as a Back-Cover Text, to the end of the list of Cover Texts in the Modified Version. Only one passage of Front-Cover Text and one of Back-Cover Text may be added by (or through arrangements made by) any one entity. If the Document already includes a cover text for the same cover, previously added by you or by arrangement made by the same entity you are acting on behalf of, you may not add another; but you may replace the old one, on explicit permission from the previous publisher that added the old one. The author(s) and publisher(s) of the Document do not by this License give permission to use their names for publicity for or to assert or imply endorsement of any Modified Version.

5. Combining Documents

You may combine the Document with other documents released under this License, under the terms defined in section 4 above for modified versions, provided that you include in the combination all of the Invariant Sections of all of the original documents, unmodified, and list them all as Invariant Sections of your combined work in its license notice, and that you preserve all their Warranty Disclaimers. The combined work need only contain one copy of this License, and multiple identical Invariant Sections may be replaced with a single copy. If there are multiple Invariant Sections with the same name but different contents, make the title of each such section unique by adding at the end of it, in parentheses, the name of the original author or publisher of that section if known, or else a unique number. Make the same adjustment to the section titles in the list of Invariant Sections in the license notice of the combined work.

In the combination, you must combine any sections Entitled "History" in the various original documents, forming one section Entitled "History"; likewise combine any sections Entitled "Acknowledgements," and any sections Entitled "Dedications." You must delete all sections Entitled "Endorsements."

6. Collections of Documents

You may make a collection consisting of the Document and other documents released under this License, and replace the individual copies of this License in the various documents with a single copy that is included in the collection, provided that you follow the rules of this License for verbatim copying of each of the documents in all other respects. You may extract a single document from such a collection, and distribute it individually under this License, provided you insert a copy of this License into the extracted document, and follow this License in all other respects regarding verbatim copying of that document.

7. AGGREGATION WITH INDEPENDENT WORKS

A compilation of the Document or its derivatives with other separate and independent documents or works, in or on a volume of a storage or distribution medium, is called an "aggregate" if the copyright resulting from the compilation is not used to limit the legal rights of the compilation's users beyond what the individual works permit. When the Document is included in an aggregate, this License does not apply to the other works in the aggregate which are not themselves derivative works of the Document.

If the Cover Text requirement of section 3 is applicable to these copies of the Document, then if the Document is less than one half of the entire aggregate, the Document's Cover Texts may be placed on covers that bracket the Document within the aggregate, or the electronic equivalent of covers if the Document is in electronic form. Otherwise they must appear on printed covers that bracket the whole aggregate.

8. TRANSLATION

Translation is considered a kind of modification, so you may distribute translations of the Document under the terms of section 4. Replacing Invariant Sections with translations requires special permission from their copyright holders, but you may include translations of some or all Invariant Sections in addition to the original versions of these Invariant Sections. You may include a translation of this License, and all the license notices in the Document, and any Warranty Disclaimers, provided that you also include the original English version of this License and the original versions of those notices and disclaimers. In case of a disagreement between the translation and the original version of this License or a notice or disclaimer, the original version will prevail. If a section in the Document is Entitled "Acknowledgements," "Dedications," or "History," the requirement (section 4) to Preserve its Title (section 1) will typically require changing the actual title.

9. TERMINATION

You may not copy, modify, sublicense, or distribute the Document except as expressly provided for under this License. Any other attempt to copy, modify, sublicense or distribute the Document is void, and will automatically terminate your rights under this License. However, parties who have received copies, or rights, from you under this License will not have their licenses terminated so long as such parties remain in full compliance.

10. Future Revisions of This License

The Free Software Foundation may publish new, revised versions of the GNU Free Documentation License from time to time. Such new versions will be similar in spirit to the present version, but may differ in detail to address new problems or concerns. See http://www.gnu.org/copyleft/. Each version of the License is given a distinguishing version number. If the Document specifies that a particular numbered version of this

License "or any later version" applies to it, you have the option of following the terms and conditions either of that specified version or of any later version that has been published (not as a draft) by the Free Software Foundation. If the Document does not specify a version number of this License, you may choose any version ever published (not as a draft).

References

CHAPTER 1

IDP FORENSIC NEW TREND ENTERPRISE BUSINESS PLAN WILL AVOID A FINANCIAL HARDSHIP

Mario Nabliba www.idpforensic.com

CHAPTER 2

TRACKING AND MANAGING DIGITAL AGE

CHAPTER 3

THE NEW FEDERAL eDISCOVERY RULES: EXPANDING YOUR SPHERE OF INFLUENCE

By Jack Molisani, Senior Member, and Johnette Hassell, Ph.D. Jack Molisani has a degree in Computer Engineering from Tulane University in New Orleans and over 20 years experience in the computer and software industries. Jack provides eDiscovery training for Electronic Evidence Retrieval, LLC when he is not running ProSpring Technical Staffing: **www.ProspringStaffing.com**

Jack also produces LavaCon, the annual conference for advanced technical communication professionals: **www.lavacon.org**

Dr. Johnette Hassell is a recognized expert witness in the field of computer forensics with over 35 years experience in the computer and software industries. She is president of Electronic Evidence Retrieval, LLC, a computer forensics company headquartered in New Orleans, Louisiana: **www.ElectronicEvidenceRetrieval. com** Dr. Hassell is a national consultant and expert witness in case evaluation, preparation, deposition and testimony. Dr. Hassell served on the faculty of Tulane University's School of Engineering for more than 25 years.

CHAPTER 4

MANUFACTURING-BUSINESS REQUIREMENT FOR ENTREPRENEURS

Manufacturing
Process Flow Charts, Procedures and Policy Statements
Team-America(A): Mario Nabliba, Rufino Fernandez, Samuel Urban
Memorandum on Riordan Manufacturing's "SR-rm-009, Internet security in the WAN"
Introduction to Information Systems Seventh Edition James A.O'Brien

CHAPTER 5

IDP FORENSIC EQUATION AND PRAGMATIC DEFINITION MARIO NABLIBA, A FOUNDER OF INTEGRITY DATA PROTECTION FORENSIC.

www.idpforensic.com

CHAPTER 6

SOME COMPUTER TIPS

www.auntsiter.com
Apple support—http:/apple.com/support
I fixit (laptop Guides)-http://www.ifit.com/
[Guidance Inc. Conference: CEIC2008—Las Vegas]

CHAPTER 7

DEFINITION

http://w3.newsmax.com/a/aftershockb/video.cfm?PROMO_CODE=CCD5-1
http://tv.breitbart.com/breitbart-news-coverage-of-iowa-caucus/
Documentsandsettings\administrator\mydocuments\documentation.avi

---

CHAPTER 8

MS IT (MASTER OF SCIENCE IN INFORMATION TECHNOLOGY) AND IT
PROJECTS ACCOMPLISHED

http://en.wikipedia.org/wiki/Systems_development_life-cycle
Software Development Life cycle
The Agile System Development Lifecycle
Pension Benefit Guaranty Corporation—Information Technology Solutions
    Lifecycle Methodology
FSA Life Cycle Framework
HHS Enterprise Performance Life Cycle Framework
The Open Systems Development Life Cycle

CHAPTER 9

GREAT QUOTES

Abraham Lincoln
Henry Ford
Francis Baron
Herbert Hoover
General Ulysses S. Grant
Aristotle
Mario Nabliba "Crowley"

CHAPTER 10

HOW TO BEHAVE WITH INTEGRITY AND RESPECT WITH OTHER NATIONS IS TO AT LEAST LEARN THEIR CONSTITUTION OR THE LAW OF THE LAND

Retrieved from "http://en.wikibooks.org/w/index.php?title=UK_Constitution_and_Government/House_of_Stuart_and_the_Commonwealth&oldid=2376348"
Category:
UK Constitution and Government

CHAPTER 11

DIRECT QUOTES REGARDING COPYRIGHTS

http://www.gnu.org/copyleft/
by the Free Software Foundation.
External links
GNU Free Documentation License (Wikipedia article on the license)
Official GNU FDL web page
[U.S. Copyright Office · Library of Congress · 101 Independence Avenue SE · Washington, DC 20559 · www.copyright.gov circular 1 reviewed: 05/2012 Printed on recycled paper]

By: Mario Nabliba, Masters
1. Supplemental (Additional) References:
2. Chapter . . . Computer Programming www.course.com/np/concepts7/ch11
3. Introduction to Information Systems Seventh Edition James A. O'Brien

1. CMGT/555. Course Notes, August 23, 2005.
2. Shelly, Gary B.; Cashman, Thomas J.; Rosenblatt, Harry J. "System Analysis and Design," Chap. 3. 2004.

1. Chapter 4 Project Integration Management
2. Chapter 5 Project Scope Management
3. Chapter 6 Project Time Management
4. lavacan 2005 Conference Automating Documentation—Peter lubbers, Oracle
5. Chapter 7 Project Cost Management
6. Chapter 8 Project Quality Management

Updated HR Integration Project Plan and Written Report
Updated HR Integration Project Plan and Written Report
Mario Nabliba

1. Chapter 5 Development Strategies
2. Chapter 9 Systems Implementations
3. https://ecampus.phoenix.edu/secure/aapd/CIST/VOP/Business/Riordan/ Internet/IndexPort

Manufacturing
Process Flow Charts, Procedures and Policy Statements
Team-America(A)

1. Chapter 3 Computer Software www.course.com/np/concepts7/ch03
2. Chapter . . . Computer Programming www.course.com/np/concepts7/ch11
3. Introduction to Information Systems Seventh Edition James A. O'Brien Memorandum on Riordan Manufacturing's "SR-rm-009, Internet security in the WAN"

1. UOP Study Materials Chapter 1 Introduction to Security 2005
2. Chapter 2 Personal Computer Security 2005
3. Chapter 5 Network Security 2005
4. Outsourcing America (Ron Hire & Anil Hire) 2005

A+ Guide to Software Managing Maintaining and Troubleshooting.
Memorandum on Organization Manufacturing's
1. UOP Chapter 12 Wide Area and Large-Scale Networks
2. Security Guide to Network Security Fundamentals

Requirements Gathering Plan
Mario Nabliba
University of Phoenix
CMGT/555—Systems Analysis And Development
Team-America (A)
Ms. Patricia Andres
September 9,2005

Requirements Gathering Plan 3
http://wiki.answers.com/Q/How_many_times_does_the_average_human_
    heart_beats_in_one_year
http://en.wikipedia.org/wiki/Point-to-point_protocol_over_Ethernet
http://wiki.answers.com/Q/How_many_times_does_the_average_human_
    heart_beats_in_one_year#ixzz203BoWG8w

Fault Tolerance Paper
University of Phoenix
CIS Risk Management and Strategic Planning Group A
Rufino Fernandez
Samuel Urban
Mario Nabliba
July 15, 2006

Author Biography

Mario Nabliba

Born in Portuguese Guinea in 1966, the small village of Tchugue. He began his education in a Portuguese school. After the independence of Portuguese Guinea in 1974, he went to boarding school, Internato, Frantz Fanon in Bor/Bissau. He completed his primary education, high school, and Institute Normal Superior in Bissau. Mario then taught high school physics and math in Bissau and Cape Verde. Mario went to the University of Lisbon, Portugal, for his postbachelor's in physics/science with an emphasis in biophysics.

In 1995, Mario came to the United States, worked in Hollywood as a proofreader of Portuguese to English, and studied English and government at the Pasadena Community College. He graduated from the University of Phoenix with a master's degree in computer science information systems. He is certified in computer forensic science, has worked as a computer system analyst with extensive experience in electronics and PI (principal investigator accredited) by NSF in tribute of scientific work/IDP Forensic in computer science. He is an expert in project management. Mario is a PhD candidate in USC.

As a teacher and a research scientist throughout his career, he is also a certified clinical expert in nutraMetrix (Advanced Nutraceuticals).

Mario has experience and knowledge of various cultural backgrounds; he can speak:
English, Portuguese, French, and Spanish.

Mario loves to act, write, and sing. As an actor, Mario did a national commercial called Fly Trap as the principal for Wrigley Orbit Gum. He auditioned and was selected in Hollywood.

Mario is most proud of his volunteer work with the Veterans Administration of Greater Los Angeles as a staff scientist.

www.ingramcontent.com/pod-product-compliance
Lightning Source LLC
Chambersburg PA
CBHW022103170526
45157CB00004B/1467